Essential Oils
Biological and Industrial Applications

Editors

Valdir Florêncio da Veiga Jr.
Military Institute of Engineering
Rio de Janeiro, Brazil

Isiaka Ajani Ogunwande
Foresight Institute of Research and Translation
Ibadan, Oyo, Nigeria

José L. Martinez
Department of Metallurgical Engineering
Faculty of Engineering
Universidad de Santiago of Chile
Santiago, Chile

CRC Press
Taylor & Francis Group
Boca Raton London New York

CRC Press is an imprint of the
Taylor & Francis Group, an **informa** business
A SCIENCE PUBLISHERS BOOK

Cover credit: Images belong to Dr. Isiaka Ajani Ogunwande.

First edition published 2025
by CRC Press
2385 NW Executive Center Drive, Suite 320, Boca Raton FL 33431

and by CRC Press
4 Park Square, Milton Park, Abingdon, Oxon, OX14 4RN

Library of Congress Cataloging-in-Publication Data (applied for)

ISBN: 978-1-032-67222-9 (hbk)
ISBN: 978-1-032-69754-3 (pbk)
ISBN: 978-1-032-69759-8 (ebk)

DOI: 10.1201/9781032697598

Typeset in Times New Roman
by Prime Publishing Services

Preface

The intricate and nuanced world of plant essential oils stands as a testament to the profound botanical mechanisms that present several functions to the vegetal and also facilitate communication among diverse species, including insects and microorganisms. Representing the pinnacle of botanical sophistication, these oils are not only captivating in their fragrances but also serve as pivotal agents in the intricate tapestry of natural history and human existence. As we embark on this aromatic journey, our senses are enveloped in the rich scents that have become intertwined with our memories of cherished places and the comforting embrace of our grandparents' kitchens. Beyond the realms of nostalgia, essential oils emerge as versatile constituents influencing various industries and permeating the fabric of our daily lives.

This scholarly compendium unfolds into ten meticulously crafted chapters and promises a comprehensive exploration of the intricate world of essential oils. Each chapter serves as a scholarly testament to the diverse roles played by these oils, transcending their conventional association with fragrances to emerge as integral components in the realms of food science, packaging, preservation, and plant protection. From enhancing knowledge of challenging complex chemical composition to understanding their robust contribution to product preservation, this essential oil book has firmly established plant volatiles as indispensable entities in our quest for sustainability and olfactory enrichment.

What makes this exploration truly remarkable is the collaborative effort emanating from multidisciplinary groups across the globe. The narrative unfolds with contributions from the diverse ecosystems of Brazil, the vast expanses of Chile, the aromatic landscapes of Nigeria, and the captivating sceneries of the Southeast Asian Region, such as the enchanting locales of Vietnam. This global convergence creates a mosaic of insights, offering readers a diverse and innovative perspective on the subject. In traversing the intricate world of essential oils, this book becomes a repository of knowledge that transcends geographical boundaries. It is a celebration of the universal impact and applications of these extraordinary natural compounds. As we turn the pages, we encounter the state of the art, the amalgamation of traditional wisdom, cutting-edge research, and the collaborative spirit of a global community united by a shared fascination for the aromatic wonders of nature.

We invite you to journey with us through these pages, from the phytochemical intricacies of essential oils to their role as potent antimicrobial agents with several daily uses. This compendium stands as a beacon, illuminating the unexplored facets of these precious gifts from nature. As the aroma of knowledge wafts through these pages, may you find inspiration, enlightenment, and a newfound appreciation for the captivating world of essential oils. Welcome to a journey that transcends borders, linking us all through the aromatic threads of history and innovation.

Editors:
Valdir Florêncio da Veiga Jr.
Isiaka Ajani Ogunwande
José L. Martinez

Contents

Section 1

Composition of Essential Oils

1

The Chemical Compositions and Biological Potentials of Essential Oils from Medicinally Important Flora of Vietnam

Le Thi Huong[1], *Do Ngoc Dai*[2*] *and Isiaka Ajani Ogunwande*[3*]

Introduction

Vietnam is situated in the tropical monsoon region where natural conditions are favourable for the formation and development of plants, especially high-value essential oil-bearing plants (Hoang 2008). Plant essential oils are complex mixtures of natural compounds, which have use as agro-feed, industrial crops, and medicinal properties. Plants are part of our daily life and essential oils have been extracted from over 3000 different species. These essential oils have domestic, industrial and medicinal uses (Adorjan and Buchbauer 2010). Essential oils have an important role in the protection of plants and are well known for their various biological and pharmacological effects. Essential oils are generally recognized as environmentally friendly, easily biodegradable, minimally toxic to mammals and have toxicity against different pathogens and insect pests.

Although chemical compositions of essential oils differ among species, the main compounds identified in essential oils often belong to the family of terpenes, which are highly lipophilic and of low molecular weight, and thus, capable of disrupting cell membranes, causing the death of cells or inhibiting sprouting of microbes (Lu et al. 2016). The chemical components of essential oils are affected by factors

[1] Faculty of Biology, College of Education, Vinh University, 182 Le Duan, Vinh City, Nghệ An Province 4300, Vietnam.

[2] Faculty of Agriculture, Forestry and Fishery, Nghe An College of Economics, 51-Ly Tu Trong, Vinh City 4300, Nghe An Province, Vietnam.

[3] University Road, Aleku Area, Osogbo, 230271, Nigeria.

* Corresponding authors: (Do Ngoc Dai and Isiaka Ajani Ogunwande) E-mail: daidn23@gmail.com; isiakaogunwande@gmail.com

such as geographical location, environment, and the stage of maturity and method of extraction. This chemical difference is directly correlated to dissimilarities in biological activities. Essential oils and their compounds exhibit activity by firstly destroying the microbial cytoplasmic wall to enhance permeability and the passage of large protons and ions.

In this chapter, the chemical compositions and biological potentials of essential oils from some species of *Amomum, Alpinia, Cinnamomum, Curcuma, Piper, Zingiber*, amongst others grown in Vietnam are highlighted.

Collection of the Different Parts of the Plants

The leaves, rhizomes, stem barks, pseudo-stem, roots, and fruits of the different plant samples were collected from different locations in Vietnam. These includes Pù Hoạt Nature Reserve (GPS: 18°17′15″N, 105°21′39″E), Xuân Thái Commune, Ben En, National Park, Thanh Hoa (GPS Coordinates: 19°482 273 N 105°462 353E), Vu Quang National Park (GPS: 18°17′15″ N, 105°21′39″ E), Ha Tinh Province, among others, as listed in Tables 1.1–1.3. All the plant samples were duly identified and authenticated, while the ascension numbers were deposited at the notable Herbaria within the country.

Isolation of Essential Oils from the Different Parts of the Plants

The different plant samples were processed either fresh or air-dried before hydrodisilation by removing debris. The known weight of each hour of the samples was subjected to separate hydrodistillation inside a Clevenger-type apparatus according to established specification (Huong et al. 2020a-c). The essential oils which were distilled in water were collected separately by running through a tap in the receiver arm of the apparatus into clean and earlier weighed sample bottles. The oils were kept under refrigeration until the moment of analyses. In most cases, the essential oil yield were also given.

Analysis of the Essential Oils

All the essential oils were analyzed by using Gas Chromatographic analysis (GC) and Gas Chromatograph-Mass Spectrometry (GC/MS) techniques. The identification of constituents of essential oils from the GC/MS spectral was performed on the basis of comparison of Retention Indices (RI Exp.) with reference to a homologous series of *n*-alkanes, under identical experimental conditions. In some cases, co-injection with known compounds under the same GC conditions was used . The mass spectral (MS) fragmentation patterns were checked with those of other essential oils of the known composition in literature.

Antimicrobial Activity Test on the Essential Oils

In most cases, the antimicrobial study was performed by agar dilution methods and microdilution broth susceptibility assay as described in various studies. Some of the microorganisms used in the study of the antimicrobial activity of the essential oils include *Enterococcus faecalis* ATCC299212, *Staphylococcus aureus* ATCC25923,

Bacillus cereus ATCC14579, *Escherichia coli* ATCC25922, *Pseudomonas aeruginosa* ATCC27853, *Salmonella enterica* ATCC13076 and *Candida albicans* ATCC10231. The Minimum Inhibitory Concentration (MIC) and median Inhibitory Concentration (IC_{50}) values were measured by the microdilution broth susceptibility assay. Several authors have identified the factors causing the variance in Minimum Inhibitory Concentration (MIC) between studies. These are incubation conditions, culture media and the use of emulsifiers or solvents, inoculum size and end-points determination (Balouiri et al. 2016).

Evaluation of Larvicidal Activity of the Essential Oils

The larvicidal activity of the essential oils was evaluated according to an established protocol (WHO 2015) with slight modifications as described earlier. In most cases, the adult larvae of *Culex. quinquefasciatus, Aedes albopictus* and *Ae. aegypti* were used. With each experiment, EtOH (96%) was used as a negative control, while permethrin, a larvicidal drug, was used as a positive control. Mortality was recorded after 24 hr and again after 48 hr of exposure during which no nutritional supplement was added.

Chemical Compositions of the Esential Oils

The major chemical compounds present in the various essential oils are given in Tables 1.1–1.3. As expected, the studied essential oils do not possess the same chemical compositions. In addition, the compositional patterns of the essential oils differ from the data reported for other essential oils. This variation in the chemical constituents of essential oils from different plants may be possibly due to nature of each plant as well as differences in the ecological and climatic conditions at the place of origin of the plants. These factors could contribute greatly to the observed variations and differences in chemical contents and biological activity when compared with samples analyzed from other parts of the world. In addition, both intra-species and inter-species variations are observed in the essential oils. For example, (*E*)-cinnamaldehyde was known to be major compounds of several *Cinnmomum* plants analyzed from different parts of the world. However, as it was noted earlier the volatile constituents of several species of *Cinnamomum* plants already reported from Vietnam contained low amounts of (*E*)-cinnamaldehyde (Table 1.3) or none. Several chemotypic forms of essential oils of plants grown in Vietnam have been reported.

Antimicrobial Activity of the Essential Oils

Alpinia Species

The chemical compositions of essential oils from some *Alpinia* plants grown in Vietnam are given in Table 1.1. The main constituents were mainly terpenes of diverse structural patterns. There are chemical variabilities in the compositions of essential oils. The essential oils displayed a number of biological activities. In the antimicrobial study, results indicated that different essential oils showed a selective inhibition of different microorganism. The essential oil from the leaves of *A. calcicola*

Table 1.1 Chemical compositions of essential oils from some *Alpinia* and *Amomum* species from Vietnam

Species	Location	Parts	Main compounds	References
			Alpinia species	
Alpinia calciola	Pu Mat National Park	Leaves	β-pinene (14.4%), δ-cadinene (9.7%), β-caryophyllene (7.7%), and α-cadinol (7.0%)	Huong et al. 2022a
		Rhizomes	α-cadinol (19.1%), β-pinene (15.3%), δ-cadinene (10.1%), and *epi*-α-muurolol (9.4%)	
Alpinia kwangsiensis	Vu Quang National Park	Leaves, stem, and rhizomes	1,8-cineole (16.9%, 14.3% and 5.8%), terpinene-4-ol (16.0%, 13.6% and 14.3%), (*E*)-methyl cinnamate (9.4%, 15.4% and 4.6%) and β-pinene (7.6%, 9.9% and 3.1%)	Nhan et al. 2021
Alpinia globosa	Xuân Thái Commune, Ben En, National Park	Leaves	β-pinene (12.1%), α-gurjunene (10.5%) and (*Z*)-13-docosenamide (9.0%)	Dai et al. 2020a
Alpinia tonkinensis		‘’	β-pinene (33.5%), (*E*)-β-ocimene (9.6%), γ-terpinene (9.2%) and α-pinene (8.4%)	
Alpinia napoensis	Xuân Thái Commune, Ben En, National Park	Rhizomes	1,8-cineole (18.2%), γ-terpinene (12.5%), terpinene-4-ol (9.6%), β-pinene (9.2%) and fenchyl acetate (7.7%)	Huong et al. 2020a
Alpinia malaccensis	Xuân Thái Commune, Bến En, National Park, Thanh Hoa	Pseudo-stem oil	β-pinene (40.8%), τ-muurolol (10.7%), α-phellandrene (9.1%) and β-phellandrene (9.1%)	Dai et al. 2020b
		Rhizomes	β-pinene (24.3%), β-phellandrene (16.7%), benzyl salicylate (8.9%) and farnesol (8.1%)	
		Fruits	methyl cinnamate (16.5%), germacrene D (16.4%) and δ-cadinene (11.8%)	
		Leaves	β-eudesmol (33.3%), β-pinene (22.5%) and δ-cadinene (8.9%)	

Species	Location	Parts	Main compounds	References
			Amomum species	
Amomum cinnamomeum	Quang Ngai, Nghe An Province	Leaves	β-pinene (35.8%), (*E*)-cinnamaldehyde (11.5%), and caryophyllene oxide (5.9%)	Huong et al. 2021a
		Rhizomes	fenchyl acetate (13.7%), camphene (12.3%), bornyl acetate (9.7%), β-phellandrene (7.4%), and eudesma-4(15),7-dien-1-β-ol (7.3%)	
Amomum rubidum	Bidoup Nui Ba National Park, Lam Dong Province	Leaves	1,8-cineole (37.7%), δ-3-carene (19.5%) and limonene (16.3%)	Huong et al. 2021b
		Stems	δ-3-carene (21.9%), limonene (17.8%) and β-phellandrene (14.6%)	
		Rhizomes	β-phellandrene (16.1%), limonene (14.4%), and δ-3-carene (13.9%)	Huong et al. 2020b
Amomum velutinum	Vu Quang National Park	Leaves	myrtenyl acetate (10.5%) and thymol (6.8%)	Huong et al. 2020c
		Pseudo-stem	β-caryophyllene (14.0%), *cis*-β-elemene (9.8%) and α-pinene (5.2%)	
Amomum glabrum	Pù Hoạt Nature Reserve	Leaves	β-pinene (62.2%) and α-pinene (13.1%)	Chung et al. 2020
		Rhizomes	β-pinene (53.7%), α-pinene (10.1%) and fenchyl acetate (11.3%)	

Table 1.2 Chemical compositions of essential oils from some *Zingiber* species from Vietnam

Species	Location	Parts	Main compounds	References
Zingiber castaneum	Pù Hoạt Nature Reserve	Leaves	bicyclogermacrene (24.8%), germacrene D (12.9%), *cis*-β-elemene (11.2%) and β-pinene (10.3%)	Huong et al. 2020b
		Rhizomes	sabinene (22.9%) and camphene (21.2%)	
Zingiber nitens		Leaves	β-pinene (45.8%) and α-pinene (10.7%)	
		Rhizomes	terpinen-4-ol (77.9%)	
Zingiber castaneum		Pseudo-stem	bicyclogermacrene (15.8%), *cis*-β-elemene (9.8%) and germacrene D (9.2%)	Huong et al. 2020e
Zingiber collinsii	Thượng Nhật Commune, NZingiber Montamumam Đông District	Leaves	α-pinene (25.6%), β-caryophyllene (16.8%), β-pinene (16.1%) and bicyclogermacrene (6.9%)	Huong et al. 2020f
		Rhizomes	camphene (22.5%), β-pinene (16.3%), α-pinene (9.0%) and humulene oxide II (9.0%)	
Zingiber montanum	Bình Chuẩn Commune, Pù Huống, Natural Reserve	Leaves	β-pinene (13.8%), β-phellandrene (11.3%) and α-pinene (7.3%)	Huong et al. 2020g
		Rhizomes	sabinene (41.1%), terpinen-4-ol (22.7%) and (*E*)-nerolidol (14.3%)	
Zingiber zerumbet	Bến En National Park, Thanh Hóa Province	Rhizome	zerumbone (51.3%)	Huong et al. 2019b
		Leaves	β-pinene (30.8%), β-caryophylene (13.0%) and caryophyllene oxide (12.0%)	Huong et al. 2020h
		Pseudo-stem	zerumbone (64.5%)	
Zingiber magang	Ba Tõ District, Ba Nam Commune, Tây Ba Tõ Protection Forest, Subzone 451, Cao stream	Leaves	β-pinene (12.3%) and (*E*)-nerolidol (10.3%)	Huong et al. 2021c
		Rhizomes	α-selinene (10.5%), caryophyllene oxide (9.7%), and *neo*-intermedeol (7.5%)	
Zingiber tamii		Leaves	β-pinene (53.8%), and sabinene (12.3%)	
		Rhizomes	β-pinene (55.4%), and sabinene (12.5%)	

Species	Location	Parts	Main compounds	References
Zingiber vuquangensis	Nậm Nhong commune, Quế Phong district, Pù Hoạt Nature Reserve	Leaves	α-pinene (11.3%) and β-pinene (38.5%)	Huong et al. 2020i
		Rhizomes	α-pinene (9.8%) and β-pinene (45.0%)	
		Leaves	*trans*-β-elemene (5.9%), and (*E*)-caryophyllene (12.2%)	
		Stem	*trans*-β-elemene (10.0%), and (*E*)-caryophyllene (14.4%)	
Zingiber recurvatum	Tri Lễ commune, Quế Phong district, Pù Hoạt Nature Reserve	Leaves	α-pinene (16.3%) and β-pinene (71.6%)	
		Rhizomes	(*E*)-caryophyllene (11.3%), bornyl acetate (10.4%), α-humulene (6.9%), and bicyclogermacrene (5.1%)	
Zingiber ottensii		Leaves	(*E*)-caryophyllene (28.0%) and *trans*-β-elemene (17.0%), and β-pinene (17.1%)	
		Rhizomes	sabinene (21.6%), terpinen-4-ol (17.1%), zerumbone (12.5%), and β-pinene (11.7%)	
Zingiber nitens	Vũ Quang National Park	Leaf	α-zingiberene (17.4%), α-pinene (11.2%), β-sesquiphellandrene (10.1%), and (*E*)-nerolidol (10.0%)	
		Rhizomes	camphene (40.4%), bornyl acetate (14.5%), (*E*)-β-ocimene (12.7%), and α-pinene (10.5%)	
Zingiber neotruncatum	Đồng Văn commune, Quế Phong district, Pù Hoạt Nature Reserve	Rhizomes	Perillene (51.3%), neral (12.3%), and geranial (17.0%)	

Table 1.2 contd.

...Table 1.2 contd.

Species	Location	Parts	Main compounds	References
Zingiber nudicarpum	Thông Thụ commune, Quế Phong district, Pù Hoạt Nature Reserve	Leaves	α-pinene (5.0–10.9%), β-pinene (0.7–34.0%), (*E*)-caryophyllene (6.4–24.3%), α-humulene (2.1–6.4%), germacrene D (0.6–6.5%), and bicyclogermacrene (3.3–16.1%)	
Zingiber cornubracteatum	Khe Kèm water fall, Pù Mát National Park	Rhizomes	α-pinene (8.2–14.5%), β-pinene (8.8–33.1%), limonene (1.0–5.1%), 1,8-cineole (2.5–10.4%), and linalool (0.4–31.0%)	
	Xuân Lý commune, Như Thanh District, Bến En National Park	Leaves	α-pinene (2.7–10.1%) and β-pinene (18.8–67.3%)	
	Tri Lễ commune, Quế Phong District, Pù Hoạt Nature Reserve	Leaves	(*E*)-caryophyllene (1.8–13.9%), germacrene D (0.7–13.7%), bicyclogermacrene (2.7–18.9%), (*E*)-nerolidol (0.9–23.0%)	

Table 1.3 Chemical compositions of essential oils from some species of flora grown in Vietnam

Species	Location	Parts	Main compounds	References
			Curcuma species	
Curcuma pambrosima	Hoa Dong Commune, Dong Hoa District, Phu Yen Province	Rhizomes	β-caryophyllene (15.4%), isoborneol (12.1%), camphor (9.7%), and β-pinene (9.5%)	Dai et al. 2022
Curcuma sahuynhensis	Quang Ngãi Province	Rhizomes	β-pinene (52.7%), β-caryophyllene (11.1%), α-pinene (8.6%), caryophyllene oxide (6.5%), and (*Z*)-β-farnesene (5.9%)	Sam et al. 2020
			Cinnamomum species	
Cinnamomum tetragonum	Pu Huong Natural Reserve	Stem bark	*cis*-geraniol (65.1%) and geranyl acetate (14.8%)	Dai et al. 2022c
		Leaves	(*E*)-cinnamaldehyde (41.0%) and *cis*-geraniol (33.8%)	
Cinnamomum doederleinii var. *raoanensis*		Leaves	linalool (41.7%), (*E*)-nerolidol (12.4%), and caryophyllene oxide (8.5%)	Dai et al. 2022d
Cinnamomum scalarinervium		Leaves	sabinene (26.6%), α-pinene (8.8%), artemisia triene (7.1%), bicyclogermacrene (6.6%), β-pinene (6.1%), and γ-terpinene (5.0%)	
			Litsea species	
Litsea umbellata	Pu Huong Natural Reserve	Leaves	β-pinene (18.8%), β-caryophyllene (16.2%), α-pinene (10.4%) and germacrene D (9.1%)	Dai et al. 2020e
Litsea iteodaphne		Leaves	β-pinene (6.6%), β-caryophyllene (21.4%), α-pinene (8.7%) and germacrene D (15.5%)	
			Vernonia species	
Vernonia solanifolia	Pu Huong Natural Reserve	Leaves	α-pinene (49.2%), δ-cadinene (10.9%), β-caryophyllene (7.8%), α-pinene (4.0%)	Huong et al. 2022c
Vernonia patula		Leaves	β-caryophyllene (28.5%), caryophyllene oxide (16.6%), α-copaene (9.0%), and α-humulene (7.1%)	Hoi et al. 2021

Table 1.3 contd.

...Table 1.3 contd.

Species	**Location**	**Parts**	**Main compounds**	**References**
			Piper species	
Piper albispicum	Kẻ Gỗ Natural Reserve	Leaves	chavicol acetate (16.3%), bicyclogermacrene (13.8%), sabinene (8.6%), β-pinene (7.7%) and α-pinene (7.3%)	Le et al. 2022
		Stem barks	β-pinene (15.6%), α-pinene (14.6%) and bicyclogermacrene (13.2%)	
Piper lolot	Dong Thap Province	Leaves	(*E*)-nerolidol (18.43%), germacrene D (12.52%), β-elemene (6.7%)	Phong et al. 2022
Piper sarmentosum	Provinces of Qu£ng Trà	Leaves	β-bisabolene (40.3%)	Huong et al. 2019c
Piper mutabile		Leaves	β-bisabolene (12.9%)	
Piper montium		Leaves	β-bisabolene (22.1%)	
Piper rubrum		Leaves	decanal (31.6%)	
Piper nigrum			δ-elemene (20.4%)	
			Hedychium species	
Hedychium stenopetalum	Vũ Quang National Park	Leaves	β-pinene (52.5%), α-pinene (11.3%) and β-caryophyllene (8.4%)	Huong et al. 2020d
		Rhizomes	β-pinene (32.2%), linalool (28.5%), α-pinene (8.5%), γ-terpinene (7.5%)	
Hedychium villosum		Leaves	β-pinene (46.8%), α-pinene (21.6%), and o-cymene (5.0%)	
		Rhizomes	β-pinene (36.1%), α-pinene (14.1%), 1,8-cineole (10.7%), linalool (9.0%), and o-cymene (6.9%)	

was active against *Pseudomonas aeruginosa* ATCC27853 and *Candida albicans* ATCC10231, with minimum inhibitory concentration (MIC) values of 5.67 μg/mL and 32.89 μg/mL, respectively. The rhizomes oil displayed the highest antimicrobial activity against *Enterococcus faecalis* ATCC299212 (MIC, 5.67 μg/mL), *Staphylococcus aureus* ATCC25923 (MIC, 5.67 μg/mL) and *Bacillus cereus* ATCC14579 (MIC, 7.67 μg/mL), respectively (Huong et al. 2022a). A study found that the rhizome essential oil of *A. kwangsiensis* was the most active against *E. faecalis* ATCC299212 with MIC value of 1.60 μg/mL, while the leaf oil displayed pronounced activity towards *B. cereus* ATCC14579 (MIC, 3.20 μg/mL). All the essential oils inhibited the growth of *S. aureus* ATCC25923 and *C. albicans* ATCC10231, with MIC value of 6.40 μg/mL (Nhan et al. 2021). The essential oil of *A. tonkinensis* inhibited the growth of *Saccharomyces cerevisiae* ATCC16404 with MIC value of 25.0 μg/mL. In addition, both essential oils of *A. globosa* and *A. tonkinensis* displayed antimicrobial activity against *Escherichia coli* ATCC25922, *S. aureus* subsp. *aureus* ATCC11632 and *Fusarium oxysporum* ATCC48112 with MIC value of 50.0 μg/mL (Dai et al. 2020a). The essential oil from the rhizome of *A. napoensis* showed stronger inhibitory effects on *E. coli, F. oxysporum,* and *S. cerevisiae* with MIC of 12.5, 27.0, and 35.0 μg/mL respectively. However, the oil displayed antimicrobial activity against *P. aeruginosa, B. subtilis,* and *Aspergilus niger* with MIC of 50 μg/mL. The oil was not active against *S. aerues* subsp. *aereus* and *C. albicans* (Huong et al. 2020a). The essential oils of *A. malaccensis* exhibited a potent and varying activity towards the tested microorganisms with MIC vaues $\leq$ 50.0 μg/mL (Dai et al. 2020b). The pseudo-stem and fruits of *A. malaccensis* inhibited the growth of *E. coli* (ATCC25922; MIC, 50 μg/mL) and *S. cerevisiae* (ATCC16404; MIC $<$ 50 μg/mL). Only the fruit essential oil displayed microbial action against *P. aeruginosa* (ATCC25923; MIC $<$ 50 μg/mL). Essential oils from fruits and rhizome exhibited antimicrobial activity on *S. aureus* subsp. *aureus* (ATCC11632; MIC $<$ 50 μg/mL). The leaf, pseudo-stem and fruit oils also showed activity against *A. niger* (ATCC9763; MIC, 50 μg/mL). Only the rhizome oil displayed potential antimicrobial action against *F. oxysporum* (ATCC48112; MIC, 50 μg/mL). However, all the oil samples did not inhibited the growth of *B. subtilis* (ATCC11774) and *C. albicans* (ATCC10231).

Amomum Species

The leaves oil of *A. cinnamomeum* displayed antimicrobial activity towards *E. faecalis* ATCC299212 (MIC, 16.0 μg/mL), *S. aureus* ATCC25923 (MIC 16.0 μg/mL), and *Bacillus cereus* ATCC14579 (MIC, 16.0 μg/mL). The oil also inhibited the growth of *C. albicans* ATCC10231, with MIC of 64.0 μg/mL. However, the rhizome oil exhibited antimicrobial activity against the four microorganisms with MIC value of 32.0 μg/mL (Huong et al. 2021a). The essential oils from the leaves of *A. rubidum* displayed stronger inhibition of *P. aeruginosa* with MIC of 25 μg/mL. At the same time the stem bark essential oil was active against *P. aeruginosa* and *C. albicans* (MIC, 50 μg/mL). Both essential oil inhibited the growth of *F. oxysporum* with MIC of 50 μg/mL. MIC values $>$ 50 μg/mL were recorded for other tested microorganisms (Huong et al. 2021b). *A. rubidum* rhizome essential oil displayed inhibitory action against *Escherichia coli* (ATCC25922) and *Fusarium oxysporum* (ATCC48112) with minimum inhibitory concentration (MIC) values of 50 μg/mL (Huong et al. 2019a). The antimicrobial study on *A. glabrum* showed that the leaf essential oil was active against *Enterococcus faecalis* ATCC299212, *Bacillus cereus* ATCC14579 and *Candida albicans* ATCC10231 with minimum inhibitory concentration (MIC) values of 4.23 μg/mL, 67.98 μg/mL and 1.56 μg/mL respectively, while MIC values of 18.67 μg/mL, 9.78 μg/mL and 10.23 μg/mL, respectively were shown by the rhizome essential oil to the same microorganisms. Both essential oils inhibited the growth of *S. aureus* ATCC25923 with the MIC value of 5.67 μg/mL (Chung et al. 2020). The essential oils from the leaves and pseudo-stem of *A. velutinum* (Huong et al. 2020c) inhibited the growth of gram-positive bacteria, namely *E. faecalis* ATCC299212 with MIC, values of 3.98–25.56 μg/mL), *S. aureus* ATCC25923 (MIC, 32.87–100.56 μg/mL), *B. cereus* ATCC14579 (MIC, 19.78–20.45 μg/mL) and a strain of yeast, *C. albicans* ATCC10231 (MIC, 20.56–55.67 μg/mL).

Zingiber Species

All the studied essential oils of the leaves and rhizomes of *Z. castaneum* and *Z. nitens* inhibited the growth of *Pseudomonas aeruginosa* ATCC25923 with MIC value of 50.0 μg/mL (Huong et al. 2022b). The pseudo-stem oil of *Z. castaneum* displayed antimicrobial activity against *P. aeruginosa* ATCC25923, *A. niger* ATCC9763 and *F. oxysporum* ATCC48112 with MIC values of 12.5 ± 0.57 μg/mL, 50 ± 1.00 μg/mL and 50 ± 0.50 μg/mL, respectively (Huong et al. 2020e). The result of the antimicrobial test indicated that *Z. zerumbet* rhizome oil inhibited the growth of *Aspergillus niger* (ATCC9763) with MIC of 50.0 μg/mL (Huong et al. 2019b). The data of antimicrobial testing indicated that *Z. zerumbet* leaf oil exhibited potent antimicrobial action against *A. niger* ATCC9763 with MIC value of 25 μg/mL. The pseudo-stem oil only displayed antimicrobial activity towards *B. subtilis* ATCC11774 with MIC of 50.0 μg/mL (Huong et al. 2020h). The rhizome essential oil of *Z. magang* displayed promising antimicrobial activity against *E. faecalis* ATCC299212 and *S. aureus* ATCC25923 with MIC values of 9.99 μg/mL and 9.67 μg/mL, respectively (Huong et al. 2021c). In addition, the essential oil from the leaf of *Z. tamii* was the most active against *E. coli* ATCC25922 (MIC, 44.38 μg/mL) and *C. albicans* ATCC10231 (MIC, 45.62 μg/mL) (Huong et al. 2021c). The *Z. nudicarpum* rhizome essential oil from Pù Hoạt Nature Reserve showed antimicrobial activity with MIC < 10 μg/mL against all three Gram-positive organisms of *E. faecalis, S. aurues* and *B. subtilis;* and MIC = 16 μg/mL against *P. aeruginosa* and *C. albicans* (Huong et al. 2020i). The leaf oil of *Z. vuquangensis, Z. recurvatum* leaf, and *Z. recurvatum* rhizome displayed activity towards *E. faecalis* with MIC = 16 μg/mL, while *Z. ottensii* rhizome had MIC = 16 μg/mL against *E. faecalis* and *C. albicans*.

Curcuma Species

The rhizome essential oil of *C. pambrosima* displayed the strongest activity against *Candida albicans* (ATCC10231) with MIC of 16.0 μg/mL. In addition, strong antimicrobial activity was recorded against *Enterococcus faecalis* (ATCC299212) and *Bacillus cereus* (ATCC14579) with MIC of 64.0 μg/mL. However, the oil displayed moderate activity towards *Staphylococcus aureus* (ATCC25923) with MIC of 256.0 µg/mL. The median inhibitory concentrations (IC_{50}) against the tested microbes were evaluated as 20.45, 100.56, 26.78, and 6.78 µg/mL, respectively (Dai et al. 2022). *Curcuma sahuynhensis* rhizome essential oil was screened against a panel of microorganisms. The essential oil showed good antibacterial activity against the Gram-positive bacteria *E. faecalis*, *S. aureus*, and *B. cereus*, and anticandidal activity against *C. albicans* with MIC value of 64 µg/mL for each organism. The essential oil was inactive against Gram-negative bacteria *E. coli*, *P. aeruginosa*, and *S. enterica*, however (Sam et al. 2020).

Vernonia Species

The essential oil from the leaf of *V. solanifolia* displayed antimicrobial activity towards four of the tested microorganisms, and anticandidal activity, with the MIC values < 50 µg/mL (Huong et al. 2020c). The essential oil exhibited antimicrobial activity against *P. aeruginosa* ATCC27853 with an MIC value of 15.68 µg/mL, and anticandidal action toward *C. albicans* ATCC10231 with an MIC value of 16.33 µg/mL. These values are active twice more than MIC values of 32.20 µg/mL, 33.56 µg/mL, and 32.76 µg/mL recorded against *E. faecalis* ATCC299212, *S. aureus* ATCC25923, and *B. cereus* ATCC14579, respectively (Huong et al. 2020c). The essential oil of *V. patula* displayed antimicrobial activity against the tested Gram-positive and Gram-negative bacteria with MIC less than 50.0 µg/mL. The leaf essential oil of *V. patula* was potent towards *E. faecalis* ATCC299212 (MIC, 8.67 µg/mL), *S. aurues* (MIC, 19.34 µg/mL), *B. cereus* (MIC, 15.67 µg/mL) and *P. aeruginosa* (MIC, 23.45 µg/mL). The essential oils also showed anti-candidal properties with MIC value of 15.99 µg/mL (Hoi et al. 2021).

Piper Species

The leaf oil of *P. albispicum* displayed antimicrobial activity towards *P. aeruginosa* ATCC27853 with the MIC value of 5.82 µg/mL. The leaves and the stems of essential oils exhibited similar activities against *E. faecalis* ATCC299212 (MIC's, 9.07 µg/mL and 9.81 µg/mL, respectively) and *C. albicans* ATCC10231 (MIC's, 10.66 µg/mL and 10.91 µg/mL, respectively). However, the stem oil exhibited a pronounced activity against *S. aureus* ATCC25923 and *B. cereus* ATCC14579, with MIC values of 10.22 µg/mL and 10.44 µg/mL, respectively. However, the studied essential oils did not inhibit the growth of *E. coli* ATCC25922 and *S. enterica* ATCC13076 (Le et al. 2022). The essential oil of *P. lolot* had inhibitory effects on *B. subtilis*, *E. coli*, *Edwardsiella ictaluri*, and *Streptococcus pneumonia* (Phong et al. 2022).

Hedychium Species

The leaf essential oil of *H. stenopetalum* displayed activity towards *Enterococcus faecalis* ATCC299212 (MIC, 7.88 µg/mL) than all other tested essential oils (Huong et al. 2022d). The leaf essential oil of *H. villosum* showed antimicrobial activity against *S. aureus* ATCC25923, *B. cereus* ATCC14579 and *C. albicans* ATCC10231 with MIC values of 7.89 µg/mL, 7.66 µg/mL and 8.33 µg/mL, respectively (Huong et al. 2022d). The MIC values of 8.54 µg/mL, 8.57 µg/mL and 8.67 µg/mL, respectively, were obtained by rhizome oil. The leaf oil of *H. villosum* exhibited antimicrobial activity against *P. aeruginosa* ATCC27853 with MIC of 15.67 µg/mL. The essential oils of *H. stenopetalum* did not show antimicrobial action against *P. aeruginosa* ATCC27853. Also, both essential oils of *H. stenopetalum* and *H. villosum* did not displayed antimicrobial activity towards *E. coli* ATCC25922 and *S. enterica* ATCC13076. The studied *Hedychium* essential oils possessed pronounced activity against the Gram-positive than Gram-negative bacteria (Huong et al. 2020d).

The MIC was used to determine the minimum concentration that prevented the growth of test microbes. However, IC_{50} is the concentration that achieved 50% growth inhibition. In the present work, it may be useful in drug formulation for some applications including the use of essential oils as a component of antimicrobial packaging. The antimicrobial activities of the essential oils have been related to the different chemical compounds present in this respect. Such compounds include α- and β-pinene, linalool, 1,8-cineole which are ubiquitous monoterpenoids in conifer and other aromatic plants, and each compound was widely tested against many organisms, and cancer cell lines. The present data of antimicrobial activity were also in agreement with several reports about the relationship between monoterpene content and the antimicrobial property of essential oils especially against *S. Aureus* (Leite et al. 2007, Asakawa 2021). In addition several sesquiterpene compounds including β-caryophyllene (Dahham et al. 2015) have proven to be antimicrobial materials.

Larvicidal Activity

Alpinia Species

Rhizome oil of *A. napoensis* at a concentration of 100 µg/mL, exhibited 100% mortality towards *Aedes albopictus* and *Culex. quinquefasciatus* within 24 hr and 48 hr. The oil displayed larvicidal activity against *Ae. albopictus* with median lethal concentrations (LC_{50}) values of 56.39 µg/mL (24 hr) and 53.85 µg/mL (48 hr), as well LC_{50} values of 46.67 µg/mL (24 hr) and 38.29 µg/mL (48 hr) towards *Cx. quinquefasciatus* (Huong et al. 2020a). Rhizome essential oil of *A. rubidum* displayed larvicidal activity against *Ae. aegypti* with minimum lethal concentration, LC_{50} values of 22.85 and 22.62 µg/mL respectively at 24 hr and 48 hr, while the LC_{90} values of 31.44 µg/mL and 31.03 µg/mL were obtained respectively at 24 hr and 48 hr (Huong et al. 2020b). Leaf oil of *A. malaccensis* also exhibited 100% mortality against *Ae. albopictus* (concentration, 100 µg/mL) at both testing periods of 24 hr and 48 hr (Dai et al. 2020b). However, the oil displayed much more potent mortality (100%) against *Ae. aegypti* at a much lower concentrations of 25 µg/mL (48 hr) and 50 µg/mL (24 hr and 48 hr). On the other hand, rhizome oil achieved the same mortality rate only against *Cx. quinquefasciatus* at a concentration of 100 µg/mL over the same test periods of 24 hr and 48 hr. The leaf oil was more toxic towards *Ae. aegypti* than *Ae. albopictus*. The leaf oil *A. malaccensis* displayed significant larvicidal activity against *Ae. albopictus* with LC_{50} of 33.43 µg/mL (24 hr) and

32.52 μg/mL (48 hr). However, LC_{90} values of 73.46 μg/mL and 65.50 μg/mL were also recorded, respectively at 24 hr and 48 hr. On the other hand, the leaf essential oil also exhibited greater potent larvicidal action towards *Ae. aegypti* with LC_{50} values of 11.90 μg/mL and LC_{90} 25.65 μg/mL at 24 hr, while LC_{50} values of 9.70 μg/mL and LC_{90} of 19.42 μg/mL were obtained at 48 hr. Moreover, the rhizome oil sample also showed potential larvicides towards *Cx. quinquefasciatus* with LC_{50} values of 24.58 μg/mL and LC_{90} of 82.45 μg/mL at 24 hr. At the 48 hr test period, however, the oil exhibited larvicidal activity with LC_{50} values of 11.66 μg/mL and LC_{90} of 22.70 μg/mL (Dai et al. 2020b).

Zingiber Species

The essential oil from *Z. castaneum* rhizome oil displayed larvicidal activity against *Ae. aegypti* and *Cx. quinquefasciatus* with LC_{50} values of 121.43 and 88.86 μg /mL, respectively, at 24 hr. The leaf oil exhibited activity with LC_{50} values of 39.30 μg/mL and 84.97 μg/mL, respectively. The leaf and rhizome oils of *Z. nitens* also displayed greater larvicidal action towards *Ae. aegypti* with LC_{50} values of 17.58 μg/mL and 29.60 μg/mL, respectively. Only rhizome oil displayed toxicity against *Cx. quinquefasciatus* with the LC_{50} value of 64.18 μg/mL (Huong et al. 2022b). Rhizome oil of *Z. castaneum* exhibited 100% mortality towards *Ae. albopictus* at a concentration of 100 μg/mL after 24 hr and 48 hr test periods. In addition, the essential oil displayed larvicidal activity against *Ae. albopictus* with LC_{50} values of 49.85 μg/mL and 43.93 μg/mL at 24 hr and 48 hr, respectively (Huong et al. 2020d). Rhizome oil of *Z. collinsii* demonstrated larvicidal effects towards the fourth instant larvae of mosquito vectors (Huong et al. 2020f). The highest mortality (100%) was observed at 24 hr exposure against *Ae. albopictus* (at a concentration of 100 μg/mL) and 48 hr (at a concentration of 50 and 100 μg/mL), while the highest mortality (100%) was observed for *Cx. quinquefasciatus* at 24 hr and 48 hr at a concentration of 100 μg/mL. The 24 hr mosquito larvicidal activity of rhizome oil against *Ae. albopictus* were LC_{50} = 25.51 μg/mL; LC_{90} = 40.22 μg/mL and towards *Cx. quinquefasciatus* with LC_{50} = 50.11 μg/mL and LC_{90} = 71.53 μg/mL). However, the 48 hr larvicidal activity were LC_{50} = 20.03 μg/mL and LC_{90} = 24.51 μg/mL (*Ae. albopictus*), as well as LC_{50} = 36.18 μg/mL and LC_{90} = 55.11 μg/mL (*Cx. quinquefasciatus*). On the other hand, no appreciable mortality and larvicidal activity was observed for the leaf oils (Huong et al. 2020f).

The minimum lethal concentration (larvicidal activity) LC_{50} of the rhizome oil of *Z. montanum* at 24 hr against *Aedes albopictus* was 35.17 μg/mL, while LC_{50} values of 32.20 μg/mL and 31.12 μg/mL were obtained against *Aedes aegypti* and *Culex quinquefasciatus* respectively. At 48 hr the oil displayed larvicidal action with LC_{50} values of 23.18 μg/mL, 25.58 μg/mL and 18.99 μg/mL, respectively towards *Ae. albopictus*, *Ae. Aegypti* and *Cx. quinquefasciatus*. The leaf oil did not exhibit significant mortality and larvicidal action (Huong et al. 2020g). The essential oil from the rhizomes of *Z. zerumbet* showed mortality of 98.3% (24 hr) and 100% (48 hr) against the *Ae. albopictus* at a concentration of 100 μg/mL (Huong et al. 2019b). In the same manner, mortality of 100% was displayed against *Cx. quinquefasciatus* under the tested time and concentration. The essential oil exhibited larvicidal activity towards *Cx. quinquefasciatus* showing minimum lethal concentrations, LC_{50}, values of 33.28 µg/mL (24 hr) and 21.81 µg/mL (48 hr). The LC_{50} values of 55.75 µg/mL and 36.22 µg/mL at 24 hr and 48 hr, respectively were obtained against *Ae. albopictus* (Huong et al. 2019b).

The essential oils showing the best larvicidal activity against *Ae. aegypti* were *Z. cornubracteatum* rhizome from Bến En National Park (24-hr LC_{50} = 17.0 µg/mL) and *Z. nudicarpum* leaf essential oil from Pù Hoạt (PH) Nature Reserve (24-hr LC_{50} = 19.3 µg/mL) (Huong et al. 2020i). The *Z. neotruncatum* rhizome (24-hr LC_{50} = 34.95 µg/mL), *Z. nudicarpum* (PH) rhizome (24-hr LC_{50} = 23.44 µg/mL) (24-hr LC_{50} = 19.3 µg/mL), *Z. recurvatum* rhizome (24-hr LC_{50} = 20.90 µg/mL) and *Z. ottensii* rhizome (24-hr LC_{50} = 38.16 µg/mL) were also recorded against *Ae. aegypti*. The rhizome essential oil of *Z. cornubracteatum* also demonstrated remarkable activity against *Ae. albopictus* (24-hr LC_{50} = 12.7 µg/mL). Except for *Z. recurvatum* rhizome (24-hr LC_{50} = 45.58 µg/mL), all the other essential oils demonstrated larvcidal activity at 24-hr against *Ae. albopictus* in the range 19.79–28.05 µg/mL (Huong et al. 2020i). Both the *Z. nudicarpum* leaf essential oil and rhizome essential oil were very active against *Cx. quinquefasciatus* larvae, with LC_{50} values of 12.4 and 11.5 µg/mL, respectively (Huong et al. 2020i). The larvicidal values for other essential oils is in the range of 24.31–33.58 µg/mL.

Litsea Species

The oil of *L. umbellata* displayed larvicidal activity against *Cx. quinquefasciatus* with LC_{50} of 36.19 µg/mL at both 24 hr and 48 hr, while LC_{50} of 40.09 µg/mL and 27.33 µg/mL at 24 hr and 48 hr, respectively, were exhibited towards *Ae. albopictus*. An LC_{90} value of 54.17 µg/mL was also obtained against *Cx. quinquefasciatus* at both 24 hr and 48 hr, while the values of 68.79 and 60.49 µg/mL respectively, were recorded against *Ae. albopictus* at 24 hr and 48 hr (Dai et al. 2020e). Moreover, the oil of *L. iteodaphne* exhibited activity against *Cx. quinquefasciatus* with LC_{50} of 37.20 µg/mL (24 hr) and 20.21 µg/mL (48 hr) while LC_{90} values were 39.27 µg/mL (24 hr) and 23.78 µg/mL (48 hr). Furthermore, *L. iteodaphne* oil displayed larvicidal activity against *Ae. albopictus* with LC_{50} values of 40.04 µg/mL (24 hr) and 16.63 µg/mL (48 hr) as well as LC_{90} values of 42.40 µg/mL (24 hr) and 23.64 µg/mL (48 hr) (Dai et al. 2020e).

Hedychium Species

Hedychium stenopetalum leaf and rhizome essential oils exhibited larvcidal activity towards *C. quinquefasciatus* with LC_{50} values of 31.31 µg/mL (LC_{90}, 65.71 µg/mL) and 20.67 µg/mL (LC_{90}, 53.21 µg/mL) at 24 hr, respectively. The LC_{50} values at 48 hr were 19.63 µg/mL (LC_{90}, 56.21 µg/mL) and 14.97 µg/mL (LC_{90}, 36.55 µg/mL), respectively. On the other hand, *H. villosum* oils displayed larvcidal activity with LC_{50} values of 19.58 µg/mL (leaf, LC_{90}, 26.67 µg/mL) and 19.78 µg/mL (rhizome, LC_{90}, 27.13 µg/mL) at 24 hr, while at 48 hr, the LC_{50} values were 15.49 µg/mL (LC_{90}, 20.63 µg/mL) and 15.63 µg/mL (LC_{90}, 21.58 µg/mL), respectively (Huong et al. 2022d). The leaf and rhizome oils of *H. stenopetalum* also showed larvicidal action against *A. aegypti* with the lowest LC_{50} values of 16.33 μg/mL (LC_{90}, 49.76 μg/mL) and 19.37 μg/mL (LC_{90}, 26.19 μg/mL) at 24 hr, respectively. At 48 hr, both oils displayed lower LC_{50} values of 7.77 μg/mL (LC_{90}, 17.72 μg/mL) and 17.89 μg/mL (LC_{90}, 24.76 μg/mL), respectively (Huong et al. 2022d). However, *H. villosum* essential oils exhibited lower larvicidal action against *A. aegypti* (Huong et al. 2022d) depicted higher LC_{50} values of 63.82 μg/mL (leaf, LC_{90}, 184.88 μg/mL) and 61.98 μg/mL (rhizome, LC_{90}, 173.14 μg/mL) at 24 hr, when compared with *H. stenoepatlum* oils. The LC_{50} values of 43.64 μg/mL (LC_{90}, 127.03 μg/mL) and 53.86 μg/mL (LC_{90}, 181.19 μg/mL) were obtained at 48 hr, respectively, for the leaf and rhizome oils (Huong et al. 2022d).

Piper Species

The essential oils from *Piper* species showed outstanding larvicidal activity against Ae. aegypti. The oils of *P. caninum, P. longum, P. montium*, and *P. mutabile*, had LC_{50} and LC_{90} values less than 10 μg/mL. In addition, *P. nigrum* essential oil had $LC_{50} < 10$ μg/mL and $LC_{90} < 20$ μg/mL. Although *P. cambodianum* had LC_{50} less than 10 μg/mL, the LC_{90} values were greater than 30 μg/mL (Phong et al. 2022).

The larvicial activities of essential oils have been attributed mainly to the several compounds identified in those samples. (*E*)-Caryophyllene, caryophyllene oxide, and β-pinene have shown moderate mosquito larvicidal activities (Hung et al. 2019). In addition, α-pinene, β-pinene (Lucia et al. 2007), linalool (Govindarajan, 2010), and β-caryophyllene (Perumalsamy et al. 2009) had earlier shown *Ae. aegypti* larvicidal activities with LC_{50} of 15.4, 12.1, 38.6, and 88.3 μg/mL, respectively. Germacrene D has demonstrated notable larvicidal activity against *Ae. aegypti* and *Cx. quinquefasciatus* (LC_{50} = 18.8 and 21.3 μg/mL, respectively) (Govindarajan 2010), and bicyclogermacrene was larvicidal against *Ae. albopictus* and *Cx. tritaeniorhynchus* (LC_{50} = 11.1 and 12.5 μg/mL, respectively) (Govindarajan and Benelli 2016). β-Pinene, however, has been shown to be more active than α-pinene: (–)-β-pinene (LC_{50} = 65 μg/mL against *Cx. quinquefasciatus*) (Pavela, 2015), (LC_{50} = 15.4 μg/mL against *Ae. aegypti*); (+)-β-pinene (LC_{50} = 22.4 μg/mL against *Ae. aegypti*) (Pavela 2015).

Conclusions

The flora of Vietnam has great development potential, and its application has a broad prospect. In this chapter, the varieties, regional distribution, main chemical components contained in some plant essential oils, were described and the antimicrobial and larvicidal effects of its components were enumerated. This report also combined the application of other medicinal materials. Currently, the research and development of Vietnamese flora is still insufficient, especially in areas such as the mechanism of pharmacological action, and the development of new products. The plants scientific utilization of the plants and how to exert its greater value in combination with medicine, agronomy, and insect microbiology still need to be explored.

References

Adorjan, S. and G. Buchbauer. 2010. Biological properties of essential oils: an updated review. Flav. Fragr. J. 25(6): 407–426.

Asakawa, Y. 2021. Dietary monoterpenoids. pp. 1–124. *In*: J. Xiao, S.D. Sarker and Y. Asakawa (eds). Handbook of Dietary Phytochemicals. Springer, Singapore. https://doi.org/10.1007/978-981-13-1745-3_17-2.

Balouiri, M., M. Sadiki and S.K. Ibnsouda. 2016. Methods for *in vitro* evaluating antimicrobial activity: a review. J. Pharmaceut. Anal. 6: 71–79.

Chung, N.T., L.T. Huong, T.M. Hoi, D.N. Dai and I.A. Ogunwande. 2020. Chemical compositions of essential oils and antimicrobial activity of *Amomum glabrum* S.Q. Tong (Zingiberaceae) from Vietnam. Invest Med. Chem. Pharmacol. 3(2): 44.

Dahham, S.S., Y.M. Tabana, M.A. Iqbal, M.B.K. Ahamed, M.O. Ezzat, A.S.A. Majid, et al. 2015. The anticancer, antioxidant and antimicrobial properties of the sesquiterpene β-caryophyllene from the essential oil of *Aquilaria crassna*. Molecules. 20(7): 11808–11829.

Dai, D.N., L.T. Huong, N.H. Hung, H.V. Chinh and I.A. Ogunwande. 2020a. Antimicrobial activity and chemical constituents of essential oil from the leaves of *Alpinia globosa* and *Alpinia tonkinensis*. J. Essent. Oil Bearing Plants. 23(2): 323–330.

Dai, D.N., L.T. Huong, N.H. Hung, H.V. Chinh and I.A. Ogunwande. 2020b. Biological potentials of essential oil: Antimicrobial activity, larvicidal efficacy and chemical compositions of essential oils from *Alpinia malaccensis* (Zingiberaceae) from Vietnam. pp. 103–128. *In:* What to Know about Essential Oils. Kjeldsen. Nova Science Publisher, New York. ISBN 9781536179873.

Dai, D.N., N.T.T. Lam, N.T. Cuong, T.Q. Ngan, N.C. Truong and I.A. Ogunwande. 2020c. Compositions of essential oil of *Cinnamomum tetragonum* A. Chev. Chem. Nat. Compd. 56(3): 545–547.

Dai, D.N., L.T. Huong, D.T.M. Chau, N.T. Nhan, L.T. Tung, N.T., Thao, et al. 2020d. Essential oils of *Cinnamomum doederleinii* var. *Raoanensis* Kim Dao and *Cinnamomum scalarinervium* Kosterm (Lauraceae) from Vietnam. Chem. Nat. Compd. 56(2): 351–353.

Dai, D., N.D. Hung, N.T. Chung, L.T. Huong, N.H. Hung, N.T.H. Chuong and I.A. Ogunwande. 2020e. Chemical constituents of the essential oils from the leaves of *Litsea umbellata* and *Litsea iteodaphne* and their mosquito larvicidal activity. J. Essent Oil Bearing Plants. 23(6): 1334–1344.

Dai, D.N., L.T. Huong, L.Y. Sam and I.A. Ogunwande. 2022. Chemical compositions and antimicrobial activity of essential oil from the rhizomes of *Curcuma pambrosima* Škorničk. & N.S. Lý growing in Vietnam. Chem. Nat. Compd. 58(1): 148–150.

Govindarajan, M. 2010. Chemical composition and larvicidal activity of leaf essential oil from *Clausena nisate* (Willd.) Hook. F. ex Benth (Rutaceae) against three mosquito species. Asian Pac. J. Trop. Med. 3: 874 –877.

Govindarajan, M. and G. Benelli. 2016. Ecotoxicology and environmental safety eco-friendly larvicides from Indian plants: Effectiveness of lavandulyl acetate and bicyclogermacrene on malaria, dengue and Japanese encephalitis mosquito vectors. Ecotoxicol. Environ. Saf. 133: 395–402.

Hoang, S.V. 2008. Uses and conservation of plant species in a National Park—A case study of Ben En, Vietnam. Econ. Bot. 62: 574–593.

Hoi, T.M., N.T. Chung, L.T. Huong and I.A. Ogunwande. 2021. Studies on Asteraceae: Chemical compositions of essential oils and antimicrobial activity of the leaves of *Vernonia patula* (Dryand.) Merr. And *Grangea maderaspatana* (L.) Poir. from Vietnam. J. Essent. Oil Bearing Plants. 24(3): 500–509.

Huong, L.T., N.T. Viet, L.N. Sam, C.N. Giang, N.H. Hung, D.N. Dai, et al. 2019a. Antimicrobial activity of essential oil from the rhizomes of *Amomum rubidum* growing in Vietnam. Am. J. Essent. Oil Nat. Prod. 7(4): 11–14.

Huong, L.T., H.V. Chinh, N.T.G. An, N.T. Viet, N.H. Hung, N.T.H. Thuong, et al. 2019b. *Zingiber zerumbet* rhizome essential oil: chemical compositions, antimicrobial and mosquito larvicidal activities. Eur. J. Med. Plants. 30(4): 1–12.

Huong, L.T., N.H. Hung, D.N. Dai, T. Tai, V.T. Hien, P. Satyal, et al. 2019c. Chemical compositions and mosquito larvicidal activities of essential oils from *Piper* species growing wild in central Vietnam. Molecules. 24: 3871. doi:10.3390/molecules24213871.

Huong, L.T., D.N. Dai, H.V. Chinh and I.A. Ogunwande. 2020a. Antimicrobial activity, larvicidal efficacy and chemical composition of essential oil from the rhizome of *Alpinia napoensis*. J. Biol. Active Prod. Nat. 10(3): 220–232.

Huong, L.T., N.T. Viet, L.N. Sam, C.N. Giang, N.H. Hung, D.N. Dai, et al. 2020b. Chemical composition and larvicidal activity of essential oil from the rhizomes of *Amomum rubidum* growing in Vietnam. J. Essent. Oil Bearing Plants. 23(2): 405–413.

Huong, L.T., L.D. Linh, D.N. Dai and I.A. Ogunwande. 2020c. Chemical compositions of essential oils and antimicrobial activity of *Amomum velutinum* X.E. Ye, Škorničk. & N.H. Xia (Zingiberaceae) from Vietnam. J. Essent. Oil Bearing Plants. 23(5): 1132–1141.

Huong, L.T., T.T. Huong, N.T. Bich, N.T. Viet and I.A. Ogunwande. 2020d. Larvicidal efficacy of essential oils from the rhizomes of *Zingiber castaneum* against *Aedes albopictus*. Am. J. Essent. Oil Nat. Prod. 8(2): 23–26.

Huong, L.T., T.T. Huong, N.T. Bich, N.T. Viet, O.A. Giwa-Ajeniya and I.A. Ogunwande. 2020e. Antimicrobial efficacy and chemical constituents of pseudo-stem essential oils from *Zingiber castaneum*. Trends. Phyto. Chem. 4(2): 93–100.

Huong, L.T., T.T. Huong, N.T.T. Huong, N.H. Hung, P.T.T. Dat, N.X. Luong, et al. 2020f. Mosquito larvicidal activities of the essential oil of *Zingiber collinsii* against *Aedes albopictus* and *Culex quinquefasciatus*. J. Oleo. Sci. 69(2): 153–160.

Huong, L.T., T.T. Huong, N.T.T. Huong, N.H. Hung, P.T.T. Dat, N.X. Luong, et al. 2020g. Chemical composition and Larvicidal activity of essential oils from *Zingiber montanum* against three mosquito vectors. Bol. Latinoam. Caribe Plant Mec. Aroma. 19(6): 569–579.

Huong, L.T., H.V. Chinh, D.N. Dai, I.O. Eresanya and I.A. Ogunwande. 2020h. Antimicrobial activity and constituents of leaf and pseudo-stem essential oils of *Zingiber zerumbet*. *In:* Grace Q. Chen. (ed.) Prime Archives in Plant Sciences. Hyderabad, India: Vide Leaf. pp. 1–15.

Huong, L.T., N.T. Chung, T.T. Huong, L.N. Sam, N.H. Hung, I.A. Ogunwande, et al. 2020i. Essential oils of *Zingiber* species from Vietnam: Chemical compositions and biological activities. Plants. 9(10): 1269.

Huong, L.T., L.N. Sam, B.D. Thach, D.N. Dai and I.A. Ogunwande. 2021a. Chemical compositions of essential oils and antimicrobial activity of *Amomum cinnamomeum* (Zingiberaceae) from Vietnam. Chem. Nat. Compd. 57(3): 574–577.

Huong, L.T., N.T. Viet, L.N. Sam, C.N. Giang, N.H. Hung, D.N. Dai, et al. 2021b. Antimicrobial activity of essential oils from the leaves and stems of *Amomum rubidum* Lamxay & N.S. Lý. Bol. Latinoam Caribe Plantas. Med. Aroma. 20(1): 81–89.

Huong, L.T., L.N. Sam, D.T.M. Chau, D.N. Dai and I.A. Ogunwande. 2021c. Chemical compositions of essential oils and antimicrobial activity of *Zingiber magang* and *Zingiber tamii* from Vietnam. J. Essent. Oil Bearing Plants. 24(5): 1087–1096.

Huong, T.H., L.D. Linh, D.N. Dai and I.A. Ogunwande. 2022a. Essential oil compositions and antimicrobial activity of the leaves and rhizomes of *Alpinia calcicola* Q.B. Nguyen & M.F. Newman from Vietnam. Chem. Nat. Compd. 58(5): 939–942.

Huong, L.T., T.T. Huong, N.T. Bich, N.T. Viet and I.A. Ogunwande. 2022b. Chemical compositions, larvicidal and antimicrobial activities of *Zingiber castaneum* (Škorničk. & Q.B. Nguyễn) and *Zingiber nitens* (M.F. Newman) essential oils. Braz. J. Pharm. Sci. 58: e200204. 17 pages.

Huong, L.T., T.N. Hoi, N.T. Chung, T.D. Binh, D.H. Son and I.A. Ogunwande. 2022c. The leaf essential oil of *Vernonia solanifola* Benth: composition and antimicrobial activity. Chem. Nat. Compd. 58(5): 943–946.

Huong, L.T., T.T. Huong, D.N. Dai and I.A. Ogunwande. 2022d. Chemical compositions, mosquito larvicidal and antimicrobial activities of essential oils of *Hedychium stenopetalum* Lodd. And *Hedychium villosum* Wall. Of Zingiberaceae family from Vietnam. J. Essent. Oil Bearing Plants. 25(4): 924–938.

Hung, N.H., P. Satyal, H.V. Hieu, N.T.H. Chuong, D.N. Dai, L.T. Huong, et al. 2019. Mosquito larvicidal activity of the essential oils of Erechtites species growing wild in Vietnam. Insects. 10: 47.

Le, N.V., L.N. Sam, L.T. Huong and I.A. Ogunwande. 2022 Chemical compositions of essential oils and antimicrobial activity of *Piper albispicum* C. DC. from Vietnam. J. Essent Oil Bearing Plants. 25(1): 82–92.

Leite, AM., E.O. Lima, E.L. Souza, M. Diniz, V.N. Trajano and I.A. Medeiros. 2007. Inhibitory effect of α-pinene, β-nene and eugenol on the growth of potential infectious endocartidis causing gram-positive bacteria. Braz. J. Pharm. 43(3): 121–126.

Lu, H., C. Xu, X. Zhang, Y. Liang and X. Liu. 2016. Antibacterial effect of limonene on food-borne pathogens. J. Zhejiang Univ. (Agric. Life Sci.). 42(3): 306–312.

Lucia, A., P. Gonzalez Audino, E. Seccacini, S. Licastro, E. Zerba and H. Masuh. 2007. Larvicidal effect of *Eucalyptus grandis* essential oil and turpentine and their major components on *Aedes aegypti* larvae. J. Am. Mosq. Control Assoc. 23: 299–303.

Nhan, N.T., C.T. Lan, L.D. Linh, L.T. Huong, D.N. Dai and I.A. Ogunwande. 2021. Chemical compositions of essential oils and antimicrobial activity of *Alpinia kwangsiensis* from Vietnam. J. Essent. Oil Bearing Plants. 24(4): 714–723.

Pavela, R. 2015. Acute toxicity and synergistic and antagonistic effects of the aromatic compounds of some essential oils against *Culex quinquefasciatus* Say larvae. Parasitol. Res. 114: 3835–3853.

Perumalsamy, H., N.-J. Kim and Y.-J. Ahn. 2009. Larvicidal activity of compounds isolated from *Asarum heterotropides* against *Culex pipiens* pallens, *Aedes aegypti*, and *Ochlerotatus togoi* (Diptera: Culicidae). J. Med. Entomol. 46: 1420–1423.

Phong, H.X., N.N. Thanh, D.T.K. Tien and B.H.D. Long. 2022. Composition and bioactivity of essential oils from lolot pepper (*Piper lolot* C. DC.) extracted by microwave-assisted hydrodistillation. Food Res. 6(5): 116–123.

Sam, L.N., L.T. Huong, P.N. Minh, B.T. Vinh, D.N. Dai, W.N. Setzer, et al. 2020. Chemical composition and antimicrobial activity of the rhizome essential oil of *Curcuma sahuynhensis* from Vietnam. J. Essent. Oil Bearing Plants. 23(4): 803–809.

World Health Organisation (WHO), 2005. Guidelines for Laboratory and Field Testing of Mosquito Larvicides. WHO /CDS /WHOPES/GCDPP, Geneva, Switzerland.

2

Comparison of the Essential Oil Content of *Schinus molle* from Different Geographical Regions and Their Applications

Maité- Rodríguez- Díaz[1*], *Jessica Bravo*[2], *Jessica Manzano-García*[3], *José L. Martinez*[4,5,6], *Jorge Escobar*[7], *Onder Yumrutas*[8], *Michelle Herrera*[1] and *Mirtha Parada*[9]

Taxonomic Characteristics

Schinus molle L. Anacardiacea is a tree, often bushy, leafy, evergreen, with a very wide crown and a thick trunk, gnarled branches that are merged with hanging branches and twigs (pendula), minute and scattered with few and small hairs (puberulae), which are lost (glabradas) with age, reaching a size of 15 m tall. The leaves are alternate, regularly imparipinnate, 10 to 30 cm long, on petioles 2 to 6 cm long, the rachis are winged a little, the leaflets (sometimes, 15) 21 to 27 (seldom 41) on each leaf, sessile, and due to their opposite, sub-opposite or alternate location, their shape ranges from linear to lanceolate, and their size (not often 1.5) 2.5 to

[1] Escuela de Química y Farmacia, Facultad de Medicina, Universidad Andrés Bello, Santiago, Chile.
[2] Facultad de Medicina, Universidad Diego Portales, Santiago, Chile.
[3] Instituto de Antropología de Córdoba (IDACOR), Universidad Nacional de Córdoba, Córdoba, Argentina.
[4] Department of Metallurgical Engineering, Faculty of Engineering, Universidad de Santiago of Chile.
[5] Facultad de Farmacia y Bioquímica, Universidad Nacional de Trujillo, Trujillo, Perú.
[6] Facultad de Ciencia Biológicas, Universidad Nacional de Trujillo, Trujillo, Perú.
[7] Facultad de Ciencias, Pontificia Universidad Católica de Valparaíso, Valparaíso, Chile
[8] Faculty of Medicine, University of Adıyaman, Adıyaman, Turkey.
[9] Instituto de Salud Pública, Santiago, Chile.
* Corresponding author: maite.rodriguez@unab.cl

6 cm long and 0.3 to 1 cm wide, their apex ranges from acute to acuminate and is usually curved at the end, the margin is entire or somewhat serrated, its texture is membranous to slightly coriaceous, they are glabrous or have small and few hairs, and they also have abundant aromatic resin. The inflorescence is in paniculate form, they are axillary, although sometimes they give the appearance of terminals, 8 to 15 (not often 20) cm long, with scarce and small hairs; the bracts are deltoid, with hairs on the back and on the margin; the 2 mm long pedicels have few hairs. Small flowers, with radial symmetry, greenish-yellow to whitish, unisexual but often with rudiments of the other sex (in the female flowers, the stamens are reduced and the anthers are empty, and in the male flowers the ovary is rudimentary); cup-shaped calyx, with five ovate to semicircular lobes, about 0.5 mm long, with hairs on margin; petals five, inserted at the base of an annular disk, elliptic to oblong, about 2 mm long; 10 stamens arranged in two series, inserted in the disk, with fine filaments of different lengths, 1 to 1.5 (rarely 2) mm long, anthers oblong, about 0.8 mm long; superior ovary, tricarpelar, trilocular but with a single fertile cavity and the other two cavities are extremely reduced so that it appears as unilocular, with a hanging ovule, the styles are three, short and thick, stigmas capitate. Fruits in the form of drupes, are small and fleshy during development, dry when ripe, globose, bright red, about 5 mm in diameter, glabrous, with a single seed.

Geographical Distribution

The species *Schinus molle* L. is naturally found in the Andes region, especifically in Peru. However, it has a wide geographic distribution, found in countries such as Colombia, Ecuador, Bolivia, Chile, Argentina, Paraguay and Uruguay (Schulte et al. 1992). Due to its wide adaptability, it can be found at altitudes up to 3900 m.a.s.l., in areas with up to 700 mm of annual rainfall (Jøker et al. 2002).

The species tolerate high temperatures, for short periods, and is resistant to drought and frost. It is a fast-growing tree, found in disturbed areas with a secondary growth. It also grows without difficulty on stony soils and slopes, tolerates most types of soil, and endures different levels of salinity and alkalinity (Jøker et al. 2002). Thus, the adaptive characteristics of the species have allowed its distribution to spread through cultivation and feral breeding, that is found in North America, the Caribbean, Central America, the Middle East, Israel and the Mediterranean areas of Europe (García and Ormazabal 2008). Its invasiveness is prominent in parts of North America, Australia, and South Africa. Some of the historical data collected indicate that, in the case of South Africa, it was introduced around 1850, where it was widely planted along roadsides (Richardson et al. 2010).

The tree was taken to Mexico between 1540 and 1550 by the Spanish, where it spread by humans and birds (Corkidi et al. 1991). The presence of *S. molle* in Europe goes back to the beginning of the 18th century, probably first in Spain, from where it was dispersed by humans to other Mediterranean countries. In Chile, it was introduced from Peru during the Inca period, growing naturally in the province of Tarapacá, extending south to Santiago (Huerta et al. 2010); in this country, the fox (Pseudalopex culpaeus) is one of the species recorded as a fruit eater and therefore seed disperser (Castro et al. 1994).

Ethnobotany

The species *S. molle* besides having several synonymies also has different common names according to the area in which it is found, even varying within the same territory. Among the vernacular names are: pirul, pirú, árbol del Perú in Mexico; molle, cuyash, huaribay in Peru; aymara in Bolivia; muelle, falso pimiento, pimiento in Colombia; mulli in Ecuador; pimenteiro, aroeira, amescla, aroeira-periquita, bálsamo, fruto de sabiá in Brazil; aguaribay, terebinto, árbol de la pimienta, gualeguay in Argentina and pimiento in Chile (Luna 2012).

Similarly, the uses of the tree are diverse, however, they tend to be less intensive in countries where the tree is not native compared to its place of origin (Goldstein and Coleman 2004, Carrere 2009). In general terms, the species has a multiplicity of uses, among which are: as a aromatizer, used as chewing gum, colorant, fuel, edible, condiment, cosmetic, tannery, forage, medicinal and insecticide (Conabio 2022). The presence of molle is perceived as a well known improvement in fertility and mitigation of soil erosion, so that in countries such as Chile the cultivation of this species has been recommended to control erosion, regarded in reforestation programs in desert areas (Rodríguez-Díaz and Rebolledo 2021). The molle fulfills different roles, including: windbreak, valued as an ornament, appreciated for its shade and wildlife refuge (Conabio 2022).

Traditional Use (Part(s) Used)

The traditional uses of *Schinus molle* are ornamental, diet, agriculture, and medicine. They include effects such as antibacterial, anti-rheumatic, antiseptic, antifungal, anti-inflammatory, analgesic, antiviral, anti-tumoral, antioxidant, antispasmodic, astringent, antipyretic, cicatrizant, cytotoxic, diuretic, purgative, wounds, toothaches, fractures among others.

The essential oil in its chemical composition could justify its therapeutic use. It is important to note that essential oils and extracts of aromatic plants have been recognized for many years as a great source of pharmaceutical agents and food additives (Doleski Muhd et al. 2015).

Ornamental Value

From the ornamental point of view, being a long-lived, evergreen tree, with its evergreen and light foliage, accompanied by elegantly hanging flowers and fruits that emanate a fragrance of public interest, is associated with the beautification of the environment where it is planted (Rodríguez-Díaz and Rebolledo 2021).

S. molle has a long history of uses in folk medicine worldwide as an ornamental plant in gardens and urban landscaping since they are very resistant to pollution (Silva-Júnior et al. 2015).

It is native to South America (López et al. 2014) and was introduced and naturalized in southern Europe, including Portugal (do Rosário Martins et al. 2014). Furthermore, *S. molle* is also cultivated as an ornamental garden tree in many warmer parts of the world (Abdel-Sattar et al. 2010). For instance, in Tunisia, *Schinus molle* was introduced as an ornamental plant in the 1900s (do Rosário Martins et al. 2014).

Additionally, in Costa Rica, *S. molle* is occasionally used as an ornamental garden tree. This plant is popular for many ailments (Díaz et al. 2008). Moreover, in Uruguay, it is present as natural stands in some regions in the North of the country and is very common as an ornamental garden tree. (Rossini et al. 1996).

The successful introduction of *S. molle* in a non-native range is attributed to its high drought-and-heat tolerance, great potential to compete for nutritive resources and light, high growth rate and prolific seed production. Herbivores and pathogens avoid the plant due to numerous noxious or repellent components (Zahed et al. 2010).

Food Uses

In ancient times, Incas used *Schinus molle* fruits as spices and to prepare a typical Peruvian alcoholic drink called "chicha de molle" (Díaz et al. 2008); (Atti dos Santos et al. 2009, Silva-Júnior et al. 2015). In its native area, the berries of *S. molle* are marketed as pepper (pink peppercorn), greatly appreciated in Mediterranean cooking (Tlilia et al. 2018).

In traditional cuisine, *S. molle* fruits (berries) have replaced black pepper for prepared alcoholic drinks and beverages (Rossini et al. 1996, do Rosário Martins et al. 2014). The berries (fruits) are aromatic and possess a sweetish, spicy, and sharp pepper-like flavor (Abdel-Sattar et al. 2010).

The volatile oil is used as an adjuvant in various applications in food products because of its antimicrobial and antioxidant properties or as antiparasitic in cattle and beekeeping (Machado et al. 2018).

The edible uses of the species vary between territories, some of the most important are the consumption of the seeds as pepper or as an additive component to the same, harvesting bunches of ripe fruits, drying them in the sun, removing the peel and packaging them for later use (Carrere 2009). Molle honey is made by boiling the liquid from the ripe fruit until it takes the consistency of syrup (Ecosaf 2010). Another known preparation is in the drinkable form with the peel of the fruit, dissolved in water that becomes a refreshing and diuretic drink, which if left to ferment becomes "chicha de molle" (Carrere 2009), even forming part of cultural rituals in the offerings to the Pachamama (Dellacassa 2010).

Agricultural Uses

Schinus molle has also been used as an ornamental plant in gardens and urban landscaping because they are very resistant to pollution. *Schinus molle's* fruit and essential leaf oils show insect repellent and insecticidal activity (Ferrero et al. 2006, Ferrero et al. 2007, Abdel-Sattar et al. 2010, Deveci et al. 2010, López et al. 2014, do Prado et al. 2018). In Ethiopia, its leaves are used as natural flies repellents (Silva-Júnior et al. 2015).

Schinus molle hexanoic extract of leaves and fruits could be a repellent and insecticidal for the Chagas disease vector *Triatoma infestans* (Díaz et al. 2008).

Essential oil of leaves, obtained through the steam distillation method, can be used as an effective repellent, and acted as a hunger inhibitor in *Sitophilus oryzae* L., popularly known as weevils. It also has insecticidal and repellent activity against

Trogoderma granarium, known as insect pest of rice, and *Tribolium castaneum*, known as brown beetle (Silva-Júnior et al. 2015).

Traditional Medicinal Uses

The leaf of the tree is one of the parts with the greatest number of popular uses, including antifungal and antimicrobial activity (Gundidza 1993). Therefore, from ethnobotany, it is often used in the control of agricultural pests (Rodríguez and Egúsquiza 1996). Other uses of the leaves are external therapeutic applications, as an antiseptic for the treatment of pediculosis or internal use (infusion) as a menstrual regulator (Madaleno and Montero 2012). Baths with water from boiled leaves also serve as a local analgesic, healing and anti-inflammatory effect (Ferrero et al. 2006). The dried foliage exposed to the sun is used topically as a poultice to mitigate ailments caused by rheumatism or sciatica, swelling of the extremities, to heal wounds (Ecosaf 2010); in addition to treating liver or stomach ailments (Ministerio de Salud Chile 2009). The infusion of the leaves is prepared to treat colds (Lombardo 1969) and when used in combination with the bark is used for the treatment of bronchitis and asthma (Ministerio de Salud Chile 2009). The decoction of the fruits is used as a local antiseptic in blennorrhagia and leucorrhea (Arrillaga 1997). The fresh fruits are often used in infusion against urine retention (Ojeda and Mesa 2008) and in the prevention and cure of urogenital infections (Ministerio de Salud Chile 2009).

The resin is another of the extractions taken from the molle tree, and has anti-spasmodic tonic properties (Ojeda and Mesa 2008). Another use of the exudate is as a chewing gum (chicle), it is considered a gum strengthener and healer of mouth ulcers (Conabio 2022). The resin is also considered to have purgative properties (Viturro et al. 2010).

Others

The wood is usually used as a timber species considered to be hard and resistant in carpentry and joinery (Ecosaf 2010). It is also used as firewood and charcoal, which are popularly considered to be good quality fuels (Viturro et al. 2010). Pigments are extracted from the bark to dye hides and wool in tanneries (Conabio 2022).

Major Chemical Constituents and Bioactive Compounds

Schinus molle L. Anacardiaceae family is a large plant, reaching six feet tall, native from American tropics, Abdel-Sattar et al. (2010). At the right season, it produces a fruit-reddish with a 5 mm diameter approximately. In some countries, this specie is used as a food. In the Andes, the fruits are used for an alcoholic drink production, called "*chichi de molle*" (Díaz et al. 2008). This specie has a long history of uses in folk medicine around the world, as an ornamental plant in gardens, and in urban landscaping, since they are very resistant to pollution (Kasimala and Kasimala 2012). In Ethiopia, its leaves are used as natural repellents against flies (Abdel-Sattar et al. 2010). In Brazil, *S. molle* var. *aroeira* is also used as an astringent, diuretic, antispasmodic, and in the control of plague control, for example, *Sitophilus oryzae* (L.) (Benzi et al. 2009).

Zahed et al. (2010) reported that the major components in fruit and leaf oils of *Schinus molle* were limonene, β-phellandrene (35.9–65.4%), α-phellandrene (24.3–20.1%), myrcene (12.8–7.7%) and α-pinene (5.9–1.7%). As can be seen, the percentage of monoterpene hydrocarbons was much higher in the leaf oil (96.3%) than in flower oil (81%). However, the percentage of the sesquiterpene hydrocarbons was approximately tenfold higher in the fruit oil as when compared to the leaf oil. In both oils, the sesquiterpene hydrocarbons fraction was dominated by germacrene-D (1.8–0.3%), γ-muurolene (1.6–0.2%) and β-caryophyllene (1.2–0.3%) for fruits and leaves, respectively. The oxygenated compounds (mono and sesquiterpenes) were more abundant in the fruit oil when compared to leaf oil. Two oxygenated sesquiterpenes, τ-cadinol and elemol were found to be the main representative components of the oxygenated fraction. Both essential oils showed a dose-dependent allelopathic activity on wheat germination and radicle elongation with the leaf oil being the more phytotoxic.

Abdel-Sattar et al. (2010) studied the fruit and leaf essential oils of *S. molle* that showed insect repellent and insecticidal activity against *Trogoderma granarium* and *Tribolium castaneum*. In these oils, 65 components were identified by the GC-MS analysis. Hydrocarbons dominated the oil composition with monoterpenes occurring in the largest amounts in fruits and leaves, 80.43 and 74.84%, respectively. p-Cymene was identified as a major component in both oils. The high yield and efficacy of *S. molle* essential oil against *T. granarium* and *T. castaneum* suggest that it could provide leads for active insecticidal agents.

The leaf essential oil hydrodistilled from *Schinus molle* grown in Costa Rica was characterized in terms of its chemical composition, antioxidant activity, ability to induce cytotoxicity and the mechanism of cell death involved in the process (Díaz et al. 2008). As a result, 42 constituents, accounting for 97.2% of the total oil, were identified. The major constituents of the oil were β-pinene and α-pinene. The antioxidant activity showed an IC_{50} of 36.3 μg mL^{-1}. The essential oil was cytotoxic in several cell lines, showing that it is more effective on breast carcinoma and leukemic cell lines. The LD_{50} for cytotoxicity at 48 hr in K562 corresponded to 78.7 μg mL^{-1}, which was very similar to the LD_{50} obtained when apoptosis was measured. The essential oil did not induce significant necrosis up to 200 μg mL^{-1}, which together with the former results indicate that apoptosis is the main mechanism of toxicity induced by *S. molle* essential oil in this cell line.

Martins et al. (2014) demonstrated the dominant compounds found in the leaf and fruit Essential Oils (EOs) of *S. molle,* finding monoterpene hydrocarbons, namely α-phellandrene, β-phellandrene, β-myrcene, limonene and α-pinene. EOs showed low scavenging antioxidant activity by the DPPH free radical method and a higher activity by the β-carotene/linoleic acid method. Antimicrobial activity of EOs was observed for Gram+, Gram– pathogenic bacteria and the food spoilage fungi. EOs showed toxicity for *Artemia salina* and lower toxicity in Swiss mice.

Belhamel et al. (2008) analyzed the constituents of essential oil isolated by hydrodistillation of the aerial parts of *Schinus molle* from Ethiopia examined by GC-MS and found a total of 16 components were identified accounting for 92.2% of the oil, among them α-phellandrene (20.6%), β-Phellandrene (10.8%), α-pinene(8.7%),β-pinene(5.1%),β-myrcene(6.9%),β-elemene(5.0%),copane(6.5%), germacrene (5.8%), γ-cadinene (6.3%) and α-humulene (5.4%). The essential oil of *S. molle* showed no significant antibacterial activity against gram positive (*Staphylococcus aureus*) but less inhibition towards gram negative bacteria (*Escherichia coli*).

Essential oils were obtained by steam distillation from berries of *Schinus molle* L. and *Schinus terebinthifolius* Raddi originating from southern of Tunisia and studied by (Bendaoud et al. 2010), 57 and 62 compounds were identified in these oils, the main were α-phellandrene (46.52% and 34.38%), β-phellandrene (20.81% (Benzi et al. 2009) and 10.61% [265.15]), α-terpineol (8.38% [226.26] and 5.60% [140.03]), α-pinene (4.34% [117.29] and 6.49% [162.25]), β-pinene (4.96% [133.81] and 3.09% [77.30]) and p-cymene (2.49% [67.28] and 7.34% [183.40]), respectively. A marked quantity of γ-cadinene (18.04% [451.05]) was also identified in the *S. terebinthifolius* essential oil, whereas only traces (0.07% [1.81]) were detected in the essential oil of *S. molle*. The *in vitro* antioxidant and antiradical scavenging properties of the investigated essential oils were evaluated by using 1,1-diphenyl-2-picrylhydrazyl (DPPH) and 2,2′-Azinobis (3-ethylbenzothiazoline-6-sulfonic acid) (ABTS) assays. Essential oil of *S. terebinthifolius* showed stronger antioxidant activity in the ABTS assay, with an IC_{50} of 24 ± 0.8 mg/L, compared to *S. molle* (IC_{50} = 257 ± 10.3 mg/L). Essential oils were also evaluated for their anticancer activities against human breast cancer cells (MCF-7). *S. terebinthifolius* essential oil was more effective against tested cell lines (IC_{50} = 47 ± 9 mg/L) than that from *S. molle* (IC_{50} = 54 ± 10 mg/L).

Santos et al. (2009), reported 27 compounds that were identified for *S. molle* and 29 for *S. terebinthifolius*. The leaves and fruits of essential oils for both *S. molle* and *S. terebinthifolius* were characterized by a high percentage of sesquiterpene and monoterpene hydrocarbons. The *S. terebinthifolius* essential oils obtained from leaves and berries presented differences with those described earlier in literature, while the leave oil of *S. molle* showed a similar composition to that from Uruguay. These findings have ecological and economic significance as *S. molle* and *S. terebinthifolius* are promising species in the pharmaceutical, cosmetic, and chemical industries. Muhd et al. (2015) showed the extensive distribution of the raw material in Rio Grande do Sul state, Brazil 19 substances were separated and the major compounds were bicyclogermacrene (20.5%), betacaryophyllene (19.7%) and spathulenol (19.2%).

do Prada et al. (2019), showed a promising result of *S. molle* volatiles against the fungus *P. brasiliensis*, which causes paracoccidioidomycosis (PCM), a systemic mycosis of great clinical importance in Latin America. The major compounds found were β-pinene (25.23%), epi-α-cadinol (21.29%), α-pinene (18.72%), myrcene (11.54%) and sabinene (5.02%).

Medicinal and Pharmacological Studies

Leaves, barks, and roots are used in folk medicine for rheumatism, toothaches, menstrual disorders, pulmonary problems, urinary tract infection, high blood pressure, ulcers, tumors, anxiety, depression, and inflammation, and as an antibacterial, antiviral, antiseptic, antifungal, antioxidant, antispasmodic, analgesic, and stimulants (do Prado et al. 2019, Tlilia et al. 2018).

Essential oils of leaves and fruits of the ornamental *Schinus molle* L. species have been used for many years. Their properties include antibacterial, anti-rheumatic, antiseptic, antifungal, anti-inflammatory, analgesic, antiviral, anti-tumoral, antioxidant, antispasmodic, astringent, antipyretic, cicatrizant, cytotoxic, diuretic, purgative, wounds, toothaches, fractures. Additionally, it helps as a treatment of skin disorders, hypertension, stopping bleeding, antidepressant for menstrual disorders, bacterial infections, asthma, respiratory and urinary tract infections (Ferrero et al. 2006, Ferrero et al. 2007, Díaz et al. 2008, Atti dos Santos et al. 2009, Abdel-Sattar et al. 2010, Bendaoud et al. 2010, Deveci et al. 2010, Zahed et al. 2010, do Rosário Martins et al. 2014, López et al. 2014, Doleski Muhd et al. 2015, Silva-Júnior et al. 2015, Machado et al. 2018, do Prado et al. 2018, Tlilia et al. 2018, Morales-Rabanales et al. 2022).

Pharmacological Studies

The essential oil of *Schinus molle* is composed mainly of monoterpene hydrocarbons and some sesquiterpenes. Their major constituents are α-phellandrene, β-Phellandrene, α-pinene, β-pinene, β-myrcene, limonene, β-elemene, copane, germacrene, γ-cadinene and α-humulene for fruits and leaves. These are attributed to antibacterial, antifungal, anti-inflammatory, cytotoxic, insecticidal, phytotoxic and antioxidant activities (Díaz et al. 2008, Abdel-Sattar et al. 2010, Zahed et al. 2010, do Rosário Martins et al. 2014, Ljalem Hadush and CR 2014, Silva-Júnior et al. 2015, Chávez-Magdaleno et al. 2018, Morales-Rabanales et al. 2022). Locally, the plant is mainly used as an analgesic, antiseptic, antibacterial, purgative, and diuretic agent (Morales-Rabanales et al. 2022).

It has been demonstrated that it possesses activity against filamentous fungi such as *Fusarium moniliforme* and *Geotrichum candidum* (Chávez-Magdaleno et al. 2018). The essential oil of *S. molle* exhibits fungi toxic activity against some storage and animal pathogenic fungi (Rossini et al. 1996). Pharmacological studies have shown several properties such as sedative, anti-inflammatory, antimicrobial activity against *Staphylococcus aureus* and *Streptococcus pyogenes*, trypanocidal, repellent and insecticidal properties against *Triatoma infestans*, the vector of Chagas' disease. Biological activities have also been described for the volatile oil, such as antibacterial, antifungal, cytotoxic, and insecticidal against *Haematobia irritans* L. (Machado et al. 2018).

Phytotoxic and insect repellent properties have been reported in studies using ethanolic, petroleum ether and hexanoic extracts of leaves and fruits of *Schinus molle* (do Rosário Martins et al. 2014). The berries and the leaves contain a volatile oil that can be isolated by steam distillation (Rossini et al. 1996).

However, the biological activity of these products can be lost through the volatilization of their components or their degradation (exposure to high temperatures, oxidation and/or UV radiation), limiting the commercial applications of these oils (López et al. 2014). Some studies have reported that low concentrations of essential oil of these species progressively reduce the survival potential of insect plagues by appetite and oviposition inhibition (Silva-Júnior et al. 2015).

It is necessary to consider variations in the ecological, climatic, and geographic conditions, post-harvest drying and storage methods, extraction methods and the type of plant material used. These can interfere in the volatile oils' yield, chemical composition, and biological properties (Rossini et al. 1996).

Conclusions

The chapter of this book has deepened the importance of the essential oil of a plant that has been widely studied in scientific literature. The molle tree (*S. areira* or *S. molle*) is considered a traditional tree in many cultures of Central and South American: countries such as Brazil, Mexico, Colombia, Bolivia, Chile, Argentina and Uruguay. In this chapter we have given evidence of the various studies that exist both in agriculture, food and pharmaceuticals, which make molle a tree of our cultures.

Schinus molle L. var. areira is a native tree and its traditional ues date back to native peoples. However, today these traditions have been rescued and technological advances have supported the different uses of this tree. It is worth mentioning that we have focused on the essential oils of this plant due to its high content of terpenes, which are responsible for its various pharmacological effects, namely: antibacterial, antifungal, anti-inflammatory, cytotoxic, insecticidal, antioxidant, among others.

The rescue of our medicinal traditions regarding the native flora of our countries is part of the current work of researchers in the area of pharmacology, pharmacognosy and phytochemistry.

References

Abdel-Sattar, E., A.A. Zaitoun, M.A. Farag, S.H. El Gayed and F.M.H. Harraz. 2010. Chemical composition, insecticidal and insect repellent activity of *Schinus molle* L. leaf and fruit essential oils against *Trogoderma granarium* and *Tribolium castaneum*. Nat. Prod. Res. 24(3): 226–235.

Adriana Hoffman, J. 2010 El árbol urbano en Chile. p 202-203.

Arrillaga de Maffei, B.R. 1997. Plantas usadas en medicina natural. Montevideo: hemisferio sur. p. 156.

Atti dos Santos, A.C., M. Rossato, F. Agostini, L. Atti Serafini, P.L. dos Santos, R. Molon, et al. 2009. Chemical composition of the essential oils from leaves and fruits of *Schinus molle* L. and *Schinus terebinthifolius* Raddi from Southern Brazil. J. Essent. Oil Bear. Plants. 12(1): 16–25.

Belhamel, K., A. Abderrahim and R. Ludwig. 2008. Chemical composition and antibacterial activity of the essential oil of *Schinus molle* L. grown in Algeria. Int. J. Essent. Oil Ther. 2: 175–177.

Bendaoud, H., M. Romdhane, J.P. Souchard, S. Cazaux and J. Bouajila. 2010. Chemical composition and anticancer and antioxidant activities of *Schinus molle* L. and *Schinus terebinthifolius* Raddi berries essential oils. J. Food. Sci. 75(6): C466–472.

Benzi, V., N. Stefanazzi and A.A. Ferrero. 2009. Biological activity of essential oils from leaves and fruits of pepper tree (*Schinus molle* L.) to control rice weevil (*Sitophilus oryzae* L.). Chil. J. Agric. Res. 69: 154–159.

Carrere, R. 2009. Anacahuita (*Schinus molle*): la indígena más popular. Colección del Grupo Guayabira Sobre Especies Indígenas. 15: 1–26.

Castro, S., C. Silva, P. Meserve, J. Gutiérrez, L. Contreras and F. Jaksic. 1994. Frugivoría y dispersión de semillas de pimiento (*Schinus molle*) por el zorro culpeo (*Pseudalopex culpaeus*) en el Parque Nacional Fray Jorge (IV Región, Chile). Rev. Chil. Hist. Nat. 67: 169–176.

Chávez-Magdaleno, M.E., A.G. Luque-Alcaraz, P. Gutiérrez-Martínez, M.O. Cortez-Rocha, A. Burgos-Hernández, J. Lizardi-Mendoza, et al. 2018. Effect of chitosan-pepper tree (*Schinus molle*) essential oil biocomposites on the growth kinetics, viability and membrane integrity of colletotrichum gloeosporioides. Revista Mexicana de Ingeniería Química. 17(1): 29–45.

Conabio (Comisión Nacional para el Conocimiento y Uso de la Biodiversidad). 2022. En www.conabio.gob.mx/conocimiento/info_especies/arboles/doctos/3-anaca4m.pdf. Consulta: el 27 de enero del 2022.

Cordiki, L., S. Cacho and A. Búrquez. 1991. Dispersión del pirú (*Schinus molle* L., *Anacardiaceae*) por aves en Teotihuacán, México. Acta Botánica Mexicana. 15: 17–22.

Dellacassa, E. 2010. Normalización de productos naturales obtenidos de especies de la Flora Aromática Latinoamericana. Porto Alegre: Edipucrs. p. 337.

Deveci, O., A. Sukan, N. Tuzun and E.H. Kocabas. 2010. Chemical composition, repellent and antimicrobial activity of *Schinus molle* L. J. Med. Plants Res. 4(21): 2211–2216.

Díaz, C., S. Quesada, O. Brenes, G. Aguilar and J.F. Cicció. 2008. Chemical composition of *Schinus molle* essential oil and its cytotoxic activity on tumour cell lines. Nat. Prod. Res. (Formerly Nat. Prod. Lett.) 22(17): 1521–1534.

do Prado, A.C., H.G. Garces, E. Bagagli, V.L.M. Rall, A. Furlanetto, A. Fernandes Jr., F.B. Furtado. 2019. *Schinus molle* essential oil as a potential source of bioactive compounds: antifungal and antibacterial properties. Journal of Applied Microbiology. 126(2): 516–522. https://doi.org/10.1111/jam.14157

do Rosário Martins, M., S. Arantes, F. Candeias, M.T. Tinoco and J. Cruz-Morais. 2014. Antioxidant, antimicrobial and toxicological properties of *Schinus molle* L. essential oils. J. Ethnopharmacol. 151: 485–492.

Doleski Muhd, P.S., C.H. Ferreira Cuelho, J. Calil Brondani and M. Palermo Manfron. 2015. Chemical composition of the *Schinus molle* L. essential oil and their biological activities. Revista Cubana de Farmacia. 49(1): 132–143.

Ecosaf (Sistemas Agroforestales en áreas secas en el Departamento de Cochabamba: caracterización y descripción). 2010. La Paz, Bolivia. 150 pp

Ferrero, A.A., J.O. Werdin González and C. Sánchez Chopa. 2006. Biological activity of *Schinus molle* on *Triatoma infestans*. Fitoterapia. 77: 381–383.

Ferrero, A.A., C. Sánchez Chopa, J.O. Werdin González and R.A. Alzogaray. 2007. Repellence and toxicity of *Schinus molle* extracts on *Blattella germanica*. Fitoterapia. 78(4): 311–314.

García, N. and C. Ormazabal. 2008. Arboles Nativos de Chile. Enersis S.A. Santiago, Chile. p. 196.

Goldstein, D.J. and R.C. Coleman. 2004. *Schinus molle* L. (*Anacardiaceae*) Chicha production in the Central Andes. Econ. Bot. 58: 523–529.

Gundidza, M. 1993. Antimicrobial activity of essential oil from *Schinus molle* Linn. Central African J. Med. 39: 231–234.

Huerta, A., I. Chuffelle, K. Puga, F. Azúa and J. Araya. 2010. Toxicity and repellence of aqueous and ethanolic extracts from *Schinus molle* on elm leaf beetle *Xanthogaleruca luteola*. Crop Protection. 29: 1118–1123.

Jerzy Rzedowski y Graciela Calderón de Rzedowski, Flora del bajío y de regiones adyacentes, Anacardiaceae, Fascículo 78, septiembre de 1999, Instituto de Ecología, A.C., Centro Regional del Bajío, Pátzcuaro, Michoacán.

Jøker, D., N.T. Cruz, M.U. Morales and E. Rojas. 2002. *Schinus molle* L. Seed Leaflet. 57: 1–2.

Kasimala, M.B. and B.B Kasimala. 2012. A review on brazilian pepper plant: *Schinus molle*. Journal of Atoms and Molecules. 2(2): 6–13.

Ljalem Hadush, A. and U. CR. 2014. Chemical composition and antibactrial activity of essential oil of *Schinus molle*. Unique J. of Pharm. and Biol. Sci. 2(1): 9–12.

Lombardo, A. 1969. Árboles y arbustos. Montevideo, Nuestra Tierra. (Nuestra Tierra : 27)

López, A., S. Castro, M.J. Andina, X. Ures, B. Munguía, J.M. Llabot, et al. 2014. Insecticidal activity of microencapsulated *Schinus molle* essential oil. Ind. Crop. and Prod. 53: 209–216.

Luna, C. 2012. Distribución e importancia maderera de la familia Anacardiáceas en El Gran Chaco Argentino. Ra Ximhai. 8(3): 83–95.

Machado, C.D., V. Raman, J.U. Rehman, B.H.L.N.S. Maia, E.K. Meneghetti, V.P. Almeida, et al. 2018. *Schinus molle*: anatomy of leaves and stems, chemical composition and insecticidal activities of volatile oil against bed bug (*Cimex lectularius*). Braz. J. Pharmacogn. 29(1).

Madaleno, I.M. and M.C. Montero. 2012. El cultivo urbano de plantas medicinales, su comercialización y usos eitoterapeuticos en la ciudad de río cuarto, provincia de Córdoba, Argentina. Cuadernos Geográficos. 50: 63–85.

Martins, Mdo R., S. Arantes, F. Candeias, M.T. Tinoco and J. Cruz-Morais. 2014. Antioxidant, antimicrobial and toxicological properties of *Schinus molle* L. essential oils. J. Ethnopharmacol. 151(1): 485–492.

Ministerio de Salud Chile. 2009. Medicamentos herbarios tradicionales. 103 especies. p. 232.

Morales-Rabanales, Q.N., W.A. Coyotl-Pérez, E. Rubio-Rosas, G.S. Cortes-Ramírez, J.F. Sánchez Ramírez and N. Villa-Ruano. 2022. Antifungal properties of hybrid films containing the essential oil of *Schinus molle*: Protective effect against postharvest rot of tomato. Food Control. 134: 1–11.

Muhd, P., C. Cuelho, J. Brondani and M. Manfron. 2015. Chemical composition of the *Schinus molle* L. essential oil and their biological activities. Revista Cubana de Farmacia. 49: 1–10.

Ojeda, E. and R. Mesa. 2008. *Schinus molle* L. Gobierno de Canarias. España. p. 6.

Página web: Disponible en: http://www.conabio.gob.mx/malezasdemexico/anacardiaceae/schinus-molle/fichas/ficha.htm#:~:text=Reino%3A%20Plantae%3B%20Subreino%3A%20Traqueobionta,%3A%20Rosidae%3B%20Orden%3A%20Sapindales, consultado el: 1 de febrero de 2022.

Richardson, D.M., D.M. Iponga, N. Roura-Pascual, R.M. Krug, S.J. Milton, G.O. Hughes, et al. 2010. Accommodating scenarios of climate change and management in modelling the distribution of the invasive tree *Schinus molle* in South Africa. Ecography. 33: 1049–1061.

Rodríguez-Díaz, M. and V. Rebolledo. 2021. Descubriendo la nobleza del molle, un árbol de nuestra América. Cuadernos Botánicos Sociales. Cuadernos Médico Sociales. 61(2): 111–114.

Rodríguez, A. and R. Egusquiza. 1996. Efecto del molle (*Schinus molle*) y sus extractos en el control de *Phthorimaea operculella* en almacenes de papa. Sociedad Entomológica del Perú. Resúmenes y programa de la 38 Convención Nacional de Entomología. Chincha (Perú). SEP. FONAGRO, Chincha. p. 23.

Rossini, C., P. Menéndez, E. Dellacassa and P. Moyna. 1996. Essential oils from leaves of *Schinus molle* and *S. lentiscifolius* of Uruguayan Origin. J. Essent. Oil Res. 8(1): 71–73.

Santos, A., M. Rossato, F. Agostini, L. Serafini, P. Santos, R. Molon, et al. 2009. Chemical composition of the essential oils from leaves and fruits of *Schinus molle* L. and *Schinus terebinthifolius* Raddi from Southern Brazil. J. Essent. Oil Bear. Plants. 12: 16–25.

Schulte, A., C. Rojas and R. Rojas. 1992. Reforestación y agroforestería en Los Andes. Uso sostenido, conservación y restauración de suelos con árboles y arbustos nativos. 1. Apuntes sobre el molle (*Schinus molle* L.). Cochabamba, Bolivia: ETSFOR-FUPAGEMA-AGRUCO-ECO.

Silva-Júnior, E.F., P.G.V. Aquino, P.F.S. Santos-Júnior, I.J.S. Nascimento, E.A. Gomes, A.L.L. Silva, et al. 2015. Phytochemical compounds and pharmacological properties from *Schinus molle* Linnaeus and *Schinus terebinthifolius* Raddi (*Anacardiaceae*). J. Chem. Pharm. Res. 7(12): 389–393.

Tlilia, N., Y. Yahia, A. Feriani, A. Labidi, L. Ghazouani, N. Nasri, et al. 2018. *Schinus terebinthifolius* vs *Schinus molle*: A comparative study of the effect of species and location on the phytochemical content of fruits. Industrial Crops & Products. 122: 559–565.

Viturro, C., A. Bandoni, E. Dellacassa, L. Serafini and H. Elder. 2010. Normalización de productos naturales obtenidos de especies de la flora aromática latinoamericana—Problemática Schinus en Latinoamérica. Proyecto CYTED IV.20. R

Zahed, N., K. Hosni, N. Brahim, M. Kallel and H. Sebei (2010). Allelopathic effect of *Schinus molle* essential oils on wheat germination. Acta Physiol. Plant. 32: 1221–1227.

3

Amazonian Piperaceae Essential Oils as Sustainable Development Tools Using Pharmaceutical Technologies

Simone Braga Carneiro[1], *Ian Gardel Carvalho Barcellos Silva*[3], *Pérsia do Nascimento Abrahão*[3], *Milena Campelo Freitas de Lima*[2], *Emily Marcele Soares Silva*[2], *Ananda da Silva Antonio*[4], *Larissa Silveira Moreira Wiedermann*[2], *Ádley Antonini Neves de Lima*[5], *Letícia Scherer Koester*[6] *and Valdir Florêncio da Veiga Junior*[3*]

Introduction

Piperaceae is a botanical family with a pantropical distribution, represented by herbs, lianas, shrubs and sometimes by trees divided in five genera with approximately 2,000 species. *Piper* is the largest genus in this family and is broadly distributed in Brazil, occurring in the Atlantic Forest and Amazon region (Gogosz et al. 2012). Piperaceae species are known for their great production of essential oils, especially in their leaves. These aromatic oils present several bioactive substances that have been increasingly explored. The scientific studies of Piperaceae essential oils are focused mainly on chemical characterization, investigation of the biological properties, isolation of volatile bioactive metabolites, the description of pharmacological activities to oils and isolated substances, and further chemical modification through synthesis (Parise-Filho et al. 2011, Corral 2014, Silva et al. 2014, Branquinho et al. 2017, Oliveira et al. 2019, Salehi et al. 2019).

[1] Graduate Program in Pharmaceutical Innovation, Amazonas Federal University, Manaus-AM, Brazil.
[2] Chemistry Department, Amazonas Federal University, Manaus-AM, Brazil.
[3] Chemical Engineering Department, Military Institute of Engineering, Rio de Janeiro-RJ, Brazil.
[4] Chemistry Institute, Federal University of Rio de Janeiro, Rio de Janeiro-RJ, Brazil.
[5] Pharmacy Department, Rio Grande do Norte Federal University, Natal, RN, Brazil.
[6] Pharmacy Faculty, Federal University of Rio Grande do Sul Federal, Porto Alegre-RS, Brazil.
* Corresponding author: valdir.veiga@gmail.com

Mono and sesquiterpenes are commonly observed substances present in essential oils from several botanical origins and, also in Piperaceae. The sesquiterpene β-caryophyllene is widely applied in the pharmaceutical industries as an anti-inflammatory, anti-carcinogenic and analgesic. In addition, monoterpenes, such as linalool and eucalyptol, are largely used in perfumes and cosmetics (Fernandes et al. 2007, Sotto et al. 2010, Klauke et al. 2014, Silva et al. 2019a, Salehi et al. 2019). Interestingly, *Piper* essential oils are known for their high concentration of terpenes, but some species are unexpectedly rich in some phenylpropanoids. These remarkable substances have diverse bioactivities, a unique aroma, are slightly toxic and highly viable in the manufacturing of interesting bioproducts. Safrole and azarone, for example, are produced in high concentrations by Amazonia *Piper* species and are considered very promising as an attractive synthon to access different chemical classes of bioactive compounds, such as prostaglandin analogs, non-steroidal anti-inflammatory agents and antithrombotic compounds (Bizzo et al. 2009, Barreiro and Fraga 1999). One example of a bioactive phenylpropanoid present in Amazonian *Piper* essential oils is dillapiole, successfully tested as a fungicide (Silva et al. 2014), insecticide (Volpe et al. 2015), bactericide (Barbosa et al. 2012, Brazão et al. 2014), larvicide (Oliveira et al. 2013) and anti-inflammatory (Parise-Filho et al. 2011). Thus, dillapiole is attributed as a highly relevant metabolite in the production of bioproducts.

The application of essential oils has significantly grown as an alternative and complement to medicines in the treatment of diseases. The biological activities of the essential oils and their bioactive compounds have been targeted for potential biotechnological innovation in new pharmaceutical formulations (Bizzo et al. 2009, Raut and Karuppayil 2014, Bilia et al. 2014, Lima et al. 2016, Matos et al. 2019). Essential oils have several bioactive compounds, but they have low solubility and bioavailability due to their non-polar and lipophilic characteristics. To improve their industrial use, technological developments should be carried out to provide them with chemical and physical-chemical characteristics capable of enhancing their application in pharmaceutical formulations. Among the technological changes developed, one can highlight delivery systems such as inclusion complexes with cyclodextrins, nano-formulations and microemulsions as carrier systems for essential oils and their bioactives (Figure 3.1) (Bilia et al. 2014, Lima et al. 2016, Cheirsilp and Rakmai 2017, Basak and Guha 2017, Matos et al. 2019, Carneiro et al. 2019, Kyriakoudi et al. 2021).

The encapsulation of essential oils and their actives in micro and nano-sized systems is a strategy that provides better stability for volatile substances, in addition to providing protection against chemical degradation resulting from environmental factors. Encapsulation also increases the bioavailability and bioactivities of the compounds. Cyclodextrins is another system used to reduce the stability problems of volatile substances in essential oils and improve the bioavailability of bioactive molecules (Bilia et al. 2014, Lima et al. 2016, Carneiro et al. 2019, Matos et al. 2019, Kyriakoudi et al. 2021).

As many of Piperaceae species produce valuable essential oils with unique phytochemical composition and remarkable biological and therapeutic properties, this chapter presents a comprehensive review of the literature regarding Piperaceae

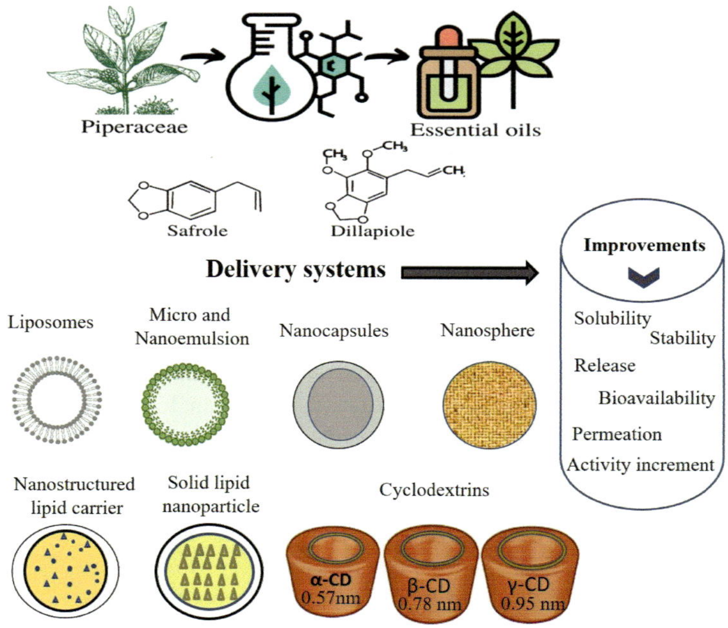

Figure 3.1 Amazonian Piperaceae essential oils and their isolated bioactives in delivery systems.

essential oils in the Amazonian region, focusing on their potential for bioproducts elaboration using pharmaceutical technologies.

Piperaceae Family

The Piperaceae family has some of the oldest tropical plants in the world, along with the Chlorantaceae family (Burger 1971), which is considered one of the most primitive angiosperm (Taylor and Hickey 1992). The Piperaceae family belongs to the Piperales order, and occurs in tropical areas around the Earth. It exists in Latin America and Malaysia (Figure 3.2) (Gogosz et al. 2012, Stevens 2012). In America, two genera have a greater occurrence: *Piper* and *Peperomia* (Cronquist, 1981). Species of Piperales have a different presentation, being seen as herbs, shrubs and arboreta, often as epiphytes or lianas (Guimarães and Valente 2001), unisexual or bisexual (Gartner 1989). The stem usually grows sympodially, presenting alternate leaves, simple, petioles and with or without stipules (Souza and Lorenzi 2005).

Many botanical families have a global distribution, though, a few of them have a rich ethnobotanical and ethnopharmaceutical history as the Piperaceae family. During several centuries in which this family has been studied as a popular medicine and food additive, its risk to human health have been well described as harmless, which enables their application in several products to human use (Tripathi et al. 1996, Parmar et al. 1997). Many Piperaceae species are used as spices and medicines in different locations, including India, Southeast Asia and Africa (Simpson and Ogorzaly 1995). The most popular product found in Piperaceae is a pepper powder produced with the fruit of *Piper nigrum* L. The black pepper has been commercialized

Figure 3.2 Geographical distribution of Piperaceae family. Adapted from Stevens, 2017.

as a seasoning over the world for many centuries, presenting a great economic importance in tropical countries, which includes India, Indonesia, Vietnam, Malaysia and Brazil (Simpson and Ogorzaly 1995).

In Brazil, there are three recognized genera, the *Manekia* Trel., the *Peperomia* Ruiz & Pav and the *Piper* L., of which the last two were classified as the genera with a greater number of species within the family (Frodin 2004). Their chemical composition and biological activities have been reviewed (Silva et al. 2017, Alves et al. 2019, Salehi et al. 2019). Nearly 500 species are found in Brazil, with main instance in the Atlantic Forest and Amazon (Souza and Lorenzi 2005). The *Piper* is considered the most relevant genus of Piperaceae, encompassing approximately 700 species, divided into herbs, shrubs, and small hanging vines, of which 266 species are typically Brazilian (Guimarães and Giordano 2004). Many Piperaceae species are used as spices and medicines in different locations, including India, Southeast Asia, and Africa (Simpson and Ogorzaly 1995). Researchers focusing on the chemical composition of *Piper* reported the presence of several bioactive compounds (Spear et al. 2005). These substances are present in plants as an important defensive mechanism against insects, nematodes and pathogenic fungi (Evans et al. 1984, Parmar et al. 1997, Lee et al. 2001).

In the Amazonian region, 100 species of Piperaceae, with predominance of *Piper* genus have been reported (Yuncker 1972). Some of these species were described as source of essential oils (Mesquita et al. 2005). In Amazonia, the *Piper* genus also has a great economic importance related with the chemical composition of their essential oils, which are already used as raw material in different industries, such as pharmaceutical, chemical and cosmetic (Silva et al. 2014). The *Piper hispidinervum* C. DC., known as long pepper, was identified as a source of an essential oil rich in safrole (Pimentel et al. 1998). Similar to the long pepper, *P. aduncum* L. (popular name: monkey pepper) has an essential oil with great economical application, that is rich in dillapiole (Parmar et al. 1997). Safrole and dillapiole are known by their large biotechnological potential, which attracted attention of several researchers, as will be further discussed.

Research with aromatic plants, especially medicinal ones, have been considered a promising approach in the search for new bioactive compounds. Besides, the traditional knowledge on the use of medicinal plants in underdevelopment countries, mainly those with tropical forests such as the Amazon, can also boost the search for bioactive species (Elisabetsky, 1991).

Chemical Composition of Essential Oils of Piperaceae

There are several papers reporting the chemical composition of essential oils extracted from Amazonian Piperaceae. The oldest record about the chemical composition of Piperaceae essential oils was published by Alencar et al. 1971, with *P. hispidinervum*. The main compounds observed in studies with one *Peperomia* and 38 *Piper* Amazonian species are listed in Table 3.1. The species *P. hispidinervum*, *P. aduncum, P. auritum*, *P. arborium* and *P. hispidum* are the most studied species of the Amazon, with the main attention in the volatile constitution of its essential oils. A published paper reviewed the chemical and biological properties of the Amazonian *Piper* essential oils (Oliveira et al. 2021b). The deep chemical variation for *Piperaceae* essential oils that allow several chemotypes and abiotic factor influencing the composition can be highlighted. On the other hand, some essential oils are so rich in some substances (more than 90.5%), especially phenylpropanoids, that can be used as pure compounds to elaborate bioproducts.

Table 3.1 Main constituents from the Amazonian Piperaceae essential oils

Specie of *Piper* (Also known as)	**Major constituent** (plant part studied)	**References**
Peperomia pellucida	Dillapiole (leaf)	Alves et al. 2019
P. acutilimbum	γ-eudesmol (leaf)	Araújo et al. 2018
P. aduncum	Dillapiole (leaf) Spathulenol (leaf) Piperitone (inflorescences)	Fazolin et al. 2007, Almeida et al. 2009, Maia et al. 1998, Silva et al. 2009, Souto et al. 2012, Potzernheim et al. 2012, Volpe et al. 2015, Majolo et al. 2019, Durofil et al. 2021, Silva et al. 2021, Valadares et al. 2018
P. amapaense	β-caryophyllene (leaf)	Santos et al. 1998
P. arboreum	Bicyclogermacrene D (leaf) Copaen-4-ol (leaf)	Mundina et al. 1998, Cysne et al. 2005, Navickiene et al. 2006, Nascimento et al. 2015 Silva et al. 2021
P. augustum	β-caryophyllene (leaf)	Noriega et al. 2019
P. auritum	Safrole (leaf)	Gupta et al. 1985, Monzote et al. 2010, Gallardo et al. 2014
P. bartlingianum	α-cadinol (leaf)	Santos et al. 1998
P. bellidifolium	(E)-nerolidol (leaf)	Araújo et al. 2018
P. cachimboense	β-caryophyllene (leaf)	Krinski et al. 2018
P. callosum	Safrole (leaf)	Santos et al. 2018, Majolo et al. 2019, Alves et al. 2021, Ayres et al. 2021
P. capitarianum	β-caryophyllene (leaf)	França et al. 2021

Specie of *Piper* (Also known as)	**Major constituent** (plant part studied)	**References**
P. carpunia	Piperitone and Limonene (leaf)	Ballesteros et al. 2019
P. claussenianum	Nerolidol and linalool (inflorescences)	Marques et al. 2015
P. consanguineum	γ-eudesmol (leaf)	Araújo et al. 2018
P. coruscans	β-caryophyllene (leaf)	Gilardoni et al. 2020
P. dilatatum	Germacrene D (aerial parts) Apiol (leaf)	Andrade et al. 2011 Colorado et al. 2019
P. divaricatum	Safrole (leaf, bark, and fruit) Eugenol and methyl eugenol (leaf)	Barbosa et al. 2012 Corpes et al. 2019, Colorado et al. 2019.
P. duckei	β-caryophyllene (leaf)	Santos et al. 1998
P. durilignum	Germacrene D (leaf)	Araújo et al. 2018
P. gaudichaudianum	Viridiflorol (leaf) Germacrene D (leaf) Biclycogermacrene, eudesmadiene, caryophyllene (leaf)	Moraes et al. 2007 Silva et al. 2021 Ramos et al. 2020
P. hispidinervum	Safrole (leaf)	Fazolin et al. 2007, Maia et al. 1987, Lima et al. 2009, Zacaroni et al. 2009, Sauter et al. 2012, Negreiros et al. 2015, Rossa et al. 2018, Santos et al. 2018, Majolo et al. 2019
P. hispidum	Dillapiole (root) Limonene (leaf) γ-terpinene (leaf)	Facundo et al. 2008 Colorado et al. 2019 Santos et al. 2018, Alves et al. 2021
P. hostmanianum	Asaricin (leaf)	Moraes et al. 2007
P. humaitanum	Caryophyllene oxide (leaf)	Moraes et al. 2007
P. jacquemontianum	Linalool, α-phellandrene and Limonene (leaf)	Zachrisson et al. 2019 Caceres et al. 2019
P. leticianum	β-caryophyllene (leaf)	Noriega et al. 2019
P. lhotzkyanum	α-phellandrene (inflorescences and infructescences)	Oliveira et al. 2021a
P. lucaeanum	α-pinene and α-zingiberene (leaf)	Marques et al. 2015
P. malacophyllom	Camphor (leaf)	Santos et al. 2012
P. marginatum	Azarone (leaf) 3,4-methylenedioxy priophenone (leaf) Biclycogermacrene (leaf) Isosafrole (leaf)	Autran et al. 2009 Alves et al. 2021 Ayres et al. 2021 Zachrisson et al. 2019

...Table 3.1 contd.

...Table 3.1 contd.

Specie of *Piper* (Also known as)	**Major constituent** (plant part studied)	**References**
P. mollicomum	1,8-cineol (leaf, dry season) Linalool and camphor (leaf, rainy season)	Ramos et al. 2020
P. nigrum	Germacrene D (leaf)	Jirovetz et al. 2002
P. oradendron	Germacrene D and pinene (leaf)	Arevalo et al. 2019
P. permucronatum	Dillapiole (leaf)	Moraes et al. 2007
P. regnellii	Germacrene D (leaf)	Anderson et al. 2018
P. sanctifelicis	3-carene (leaf)	Colorado et al. 2019
P. subscutatum	β-caryophyllene (leaf)	Ramírez et al. 2021
P. tuberculatum	β-caryophyllene (leaf and bark)	Navickiene et al. 2006, Facundo et al. 2008

Piperaceae essential oils being composed mainly by terpenes and phenylpropanoids (Figure 3.3), where the last one is considered a large but not very diversified class, metabolized though the shikimic acid pathway (Simões et al. 2007). Safrole (*P. hispidinervum, P. auritum, P. callosum* and *P. divaricatum*), isosafrole (*P. marginatum*), apiole (*P. dilatatum*) and dillapiole (*Peperomia pellucida, Piper aduncum, P. permucronatum* and *P. hispidum*) are the most usually cited, but sarisan and myristicin are frequently observed as minority compounds in essential oils and also in Piperaceae extracts.

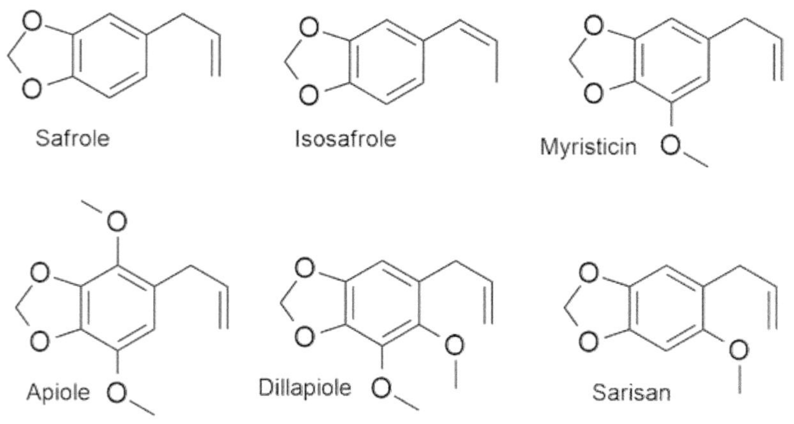

Figure 3.3 Main Phenylpropanoids of Piperaceae Essential Oil.

The main terpenoids described for Amazonia Piperaceae species belongs to monoterpene and sesquiterpenes classes (Figure 3.4). Eucalyptol (1,8-cineole), an oxygenated monoterpene, takes part in the composition of the essential oils of *P. aduncum* leaves, corresponding to the second major constituent (Oliveira et al. 2013). Since dillapiole is a very desirable constituent from *P. aduncum* essential oils, eucalyptol appears as an important byproduct. It is also the main constituent in *Piper mollicomum* (Ramos et al. 2020). Linalool, also an oxygenated monoterpene, was reported in the chemical composition of *P. aduncum* essential oils as the third major constituent, representing 30% of the composition of the leaf oils and 40% of the composition of the fruit oils, approximately (Navickiene et al. 2006).

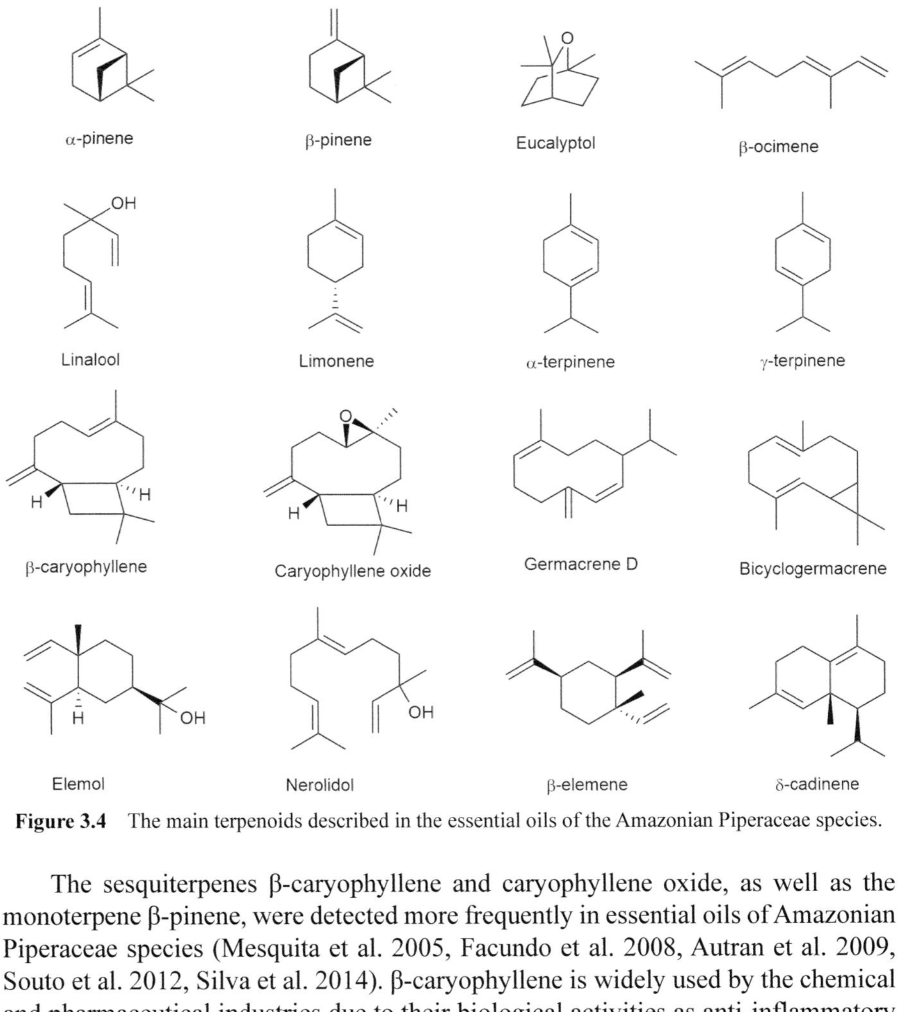

Figure 3.4 The main terpenoids described in the essential oils of the Amazonian Piperaceae species.

The sesquiterpenes β-caryophyllene and caryophyllene oxide, as well as the monoterpene β-pinene, were detected more frequently in essential oils of Amazonian Piperaceae species (Mesquita et al. 2005, Facundo et al. 2008, Autran et al. 2009, Souto et al. 2012, Silva et al. 2014). β-caryophyllene is widely used by the chemical and pharmaceutical industries due to their biological activities as anti-inflammatory (Pinheiro et al. 2017), anti-carcinogenic and analgesic (Fernandes et al. 2007, Sotto et al. 2010, Klauke et al. 2014).

Safrole is a phenylpropanoid of great economic importance, it was used as raw material in several synthetic reactions. Its main relevance is attributed to the synthesis of heliotropin (piperonal) (Grimshaw and Hua 1994), a very desirable product by the perfume industry as a fragrance fixer, and in the synthesis of piperonyl butoxide (PBO), which is a synergist pesticide. The safrole molecule has an interesting structure with regioselective functionalized carbons atoms. This enables the conversion of safrole into other molecular structures, with possible biological activities or with greater commercial applications, such as piperine and vanillin, which are used as anti-inflammatory and antioxidant, respectively (Barreiro and Fraga 1999). The species of Piperaceae family have become an alternatively sources of safrole (Bizzo et al. 2009) and other phenylpropanoids. The genetic improvement of *P. hispidinervum*, after its vegetative propagation and adaptation, have proved to be an excellent alternative applied in cultivation techniques which aims to increase the production of phenylpropanoid. With this method, the safrole content increased by 10% of the original yield, reaching about 98.0% of safrole (Maia et al. 1987, Bizzo et al. 2009). Cremasco and Braga (2010) proposed the isomerization of safrole from *P. hispidinervum* in isosafrole in the essential oil, aiming at the optimization and feasibility of heliotropin synthesis, without the need of safrole isolation.

The divergences and similarities between the species are remarkable. Briefly, *P. hispidinervum* and *P. auritum* are Amazonian species that have the highest levels of safrole (85%) (Monzote et al. 2010); most of the chemical composition of *P. aduncum* (64.4%) and *P. hispidum* (57.5%) is dillapiole, a relevant phenylpropanoid due to its excellent biological activities anti-inflammatory and antibacterial (Facundo et al. 2008, Parise-Filho et al. 2011, Brazão et al. 2014); in *P. marginatum* approximately 30% of the essential oil is composed of azarone isomers (Autran et al. 2009). Besides, it was reported that trace amounts of eugenol in *P. auritum* and methyl eugenol in *P. hispidinervum* were found (Sauter et al. 2012).

The number of chemical compositions on *Piper* species can be observed in several studies relating to the influences of abiotic factors, such as soil and climate. On the other hand, published papers studied the circadian variation on the chemical composition of essential oils obtained from branches, roots, stems, leaves, inflorescences and infructescences. In the studies performed by Ramos et al. (2020), *Piper mollicomum* was analyzed and great differences were observed considering the seasonality, vegetative part and, also phenology. The chemical variation was observed in the rainy season and the vegetative phase, changing from linalool and camphor rich oils to *E*-nerolidol and 1,8-cineol rich oils, during the day. During the dry season (reproductive phase), the oils were mainly composed by 1,8-cineole or α-terpineol. A similar study was performed by the same research group, with *Piper gaudichaudianum*, where several chemotypes were described (Ramos et al. 2021).

Despite the bioactive richness of terpenes and phenylpropanoids, these chemical classes are very nonpolar and lipophilic, hindering the development of hydrophilic formulations, more bioavailable and sensorially accepted. The volatilization of these compounds is also an issue in the formulation of stable preparations. Besides, some of the phenylpropanoids are more toxic substances and controlled release technologies are alternatives to allow a proper concentration of bioactive substances, without any undesirable effects. In this way, the association of these bioactive compounds to cyclodextrins, polymeric and lipidic nano formulations or microemulsions reduce such problems and have been considered as a more feasible strategy to develop cosmetic and pharmaceutical products (Damasceno et al. 2011, Fanun et al. 2012, Bilia et al. 2014, Lima et al. 2016, Pinheiro et al. 2017, Rakmai et al. 2017, Basak and Guha 2017, Carneiro et al. 2019, Matos et al. 2019).

Bioactive Potential of Essential Oils of the Amazon Piperaceae Species

The chemobiodiversity of Piperaceae species strengthens its biopharmacological potential where the most studied species are those of the genus *Piper* due to their ethnobotany and valuable biological activities (Parra et al. 2011). Traditional fresh leaf preparations used topically are known for their anti-inflammatory, anti-ulcer, anti-diarrheal, anti-parasitic and wound-healing properties. *In vitro* biological activities suggested, in general, the correct traditional use of leaves and their preparations (Ballesteros et al. 2019). *Piper* plants are chemically protected by a wide range of compounds. Essential oils can be used in the development of biopesticides to control weeds and pests, among other uses. Therefore, *Piper* essential oils are a potential source of natural biocidal agents for plant protection. Thus, they are presented as an alternative source for the direct use or development of herbal anthelmintics, requiring advances in studies for *in vivo* treatment (Colorado et al. 2019).

For example, studies investigate the *in vitro* efficacy of essential oils from plants of the Piperaceae family, for the control of endoparasites, namely; *Piper hispidinervum, Piper hispidum, Piper marginatum, Piper mollicomum Kunth* and *Piper callosum*. The essential oils of the studied Piperaceae showed antiparasitic efficiency of tambaqui acanthocephalons (Santos et al. 2018, Ayres et al. 2021, Cossolin et al. 2019). The insecticidal effect in the Piperaceae family has been highlighted in the fight against the dengue virus. Its use against *Aedes aegypti*, the vector responsible for the transmission of arboviruses, which compromise public health, has been an alternative to synthetic products in several studies (Souza et al. 2020, França et al. 2021).

The *P. aduncum* is the most studied species of this family concerning the biological properties of its essential oils. Despite its bactericidal and anti-inflammatory activity, dillapiole had been also considered an outstanding antifungal compound. The richness of biological activities is of great value for the intensification of studies on the insertion of dillapiole in pharmaceutical formulations as an active principle. Antifungal activities are also reported in this species mainly related to the phytopathogen *Sclerotinia sclerotiorum* which causes white mold in plants. *P. brachystachyum*, *P. falconeri*, *P. guineese* and *P. hispidum*, which belong to the Piperaceae family and are also rich in essential oils with important antifungal activities (Valadares et al. 2018).

P. hispidinervum is another Amazonian species well studied of the genus, mainly because its high content of safrole, which confers several bioactivities, such as fungicide, larvicide and insecticide. In *P. marginatum*, which has high contents of β-caryophyllene on its essential oils, is applied as a repellent, insecticidal and antifungal bioproduct. While, *P. aleyreanum*, like many other *Piper* essential oils, consists mainly of caryophyllene and its oxide. It presents antifungal, antioxidant, anti-inflammatory, antinociceptive and gastroprotection activities. The oil of *P. hispidinervum* showed great efficacy in the treatment of fungal diseases and studies show that there is a high mortality of fungi with a shorter time of exposure *in vitro* (Durofil et al. 2021, Anderson et al. 2018).

Piper lhotzkyanum Kunt is used in folk medicine in the preparation of infusions for the treatment of rheumatism, sore throats, gastrointestinal problems, bronchial problems. When chewed, they have a strong flavor and produce a burning sensation and analgesia. Essential oils are reported to have inhibitory activity against the growth of *M. tuberculosis*, as they present prototype molecules for the development of new anti-tuberculosis agents. Tuberculosis is a neglected disease caused by *Mycobacterium tuberculosis*, it is notable that the increase in multidrug-resistant drugs, further reduces the range of treatment options. In this case , there is an urgent need to discover new drugs and an alternative would be the combination of commercial drugs with natural compounds that could be tested to potentiate antibiotics (Oliveira et al. 2021). There are also results that the essential oils of *P. arboreum, P. aduncum* and *P. gaudichaudianum* have an antibacterial effect to inhibit the growth of *A. hydrophila* and *S. aureus* strains. These findings become important in the search for new effective therapies for infections triggered by multidrug-resistant bacteria (Majolo et al. 2019, Silva et al. 2021). Biological studies involving *Piper* species are being intensified, some of them are summarized at Table 3.2.

In view of the important biological activities of those Amazonian Piperaceae species, it is essential to insert the active ingredients in pharmaceutical formulations such as nanostructured systems and cyclodextrins, to provide greater stability and bioavailability, adding greater therapeutic efficacy of these bioproducts.

Delivery Strategies Applied to Essential Oils of Piperaceae

The *Piper* species presented earlier have great pharmacological importance and their insertion in formulations using cyclodextrins and nanostructured delivery systems strengthens the use of these products because they protect natural bioactive substances from environmental conditions and improve their useful life and bioavailability, potentiating their functional capacities.

The study by Rakmai et al. (2017) shows the potentiation of the activity of the essential oil of black pepper (*P. nigrum*) which has as a major substance, the β-caryophyllene after insertion in inclusion complexes with cyclodextrin HP-βCD. The application of the essential oil out of an inclusion complex presented an antibacterial activity with Minimal Inhibitory Concentration (MIC) at 1.000 $\mu g.mL^{-1}$ for *Staphylococcus aureus* and for *Escherichia coli* it presented MIC at 2.000 $\mu g.mL^{-1}$. On the other hand, the encapsulation of the essential oil in HP-βCD, enhanced the antibacterial activity of black pepper oil improving it at four times against *S. aureus*, with a MIC at 250 $\mu g.mL^{-1}$ and of 500 $\mu g.mL^{-1}$ for *E. coli*. Santos et al. (2017) also tested β-caryophyllene complexation with Mβ-cyclodextrin, observing that these complexes promoted a significant improvement in anti-inflammatory, gastric protection, and antioxidant activities in relation to pure β-caryophyllene.

Nano emulsions prepared with the essential oil of black pepper (*P. nigrum*) using ultrasound and high-pressure homogenization were also tested for their antimicrobial properties against Gram (+) and Gram (–) bacteria (Jímenez et al. 2017). The pure oil showed MICs greater than 500 $\mu L.mL^{-1}$ for *Listeria monocytogenes, S. aureus, Salmonella enterica* and *E. coli*. Nano emulsions showed MIC at 150 $\mu L.mL^{-1}$ for *L. monocytogenes, S. aureus* (MIC, 120 $\mu L.mL^{-1}$) and *S. enterica* (MIC, 160 $\mu L.mL^{-1}$).

Table 3.2 Biological activity of Amazonian Piperaceae essential oils

Specie	Biological activity	Animal and microorganism	Dose, concentration, and MIC	References
P. aleyreanum	Antifungal	*Cladosporium cladosporioides* *C. phareospermum*	DL < 0.1 μg	Silva et al. 2014
	Antioxidant	–	DPPH 36.0 ± 0.9	Silva et al. 2014
	Anti-inflammatory	Swiss mice	ID_{50} 281.2 and 70.5 mg. kg^{-1}	Lima et al. 2012
	Antinociceptive	Swiss mice	ID_{50} 53.6 and 43.5 mg.kg^{-1}	Lima et al. 2012
	Gastroprotector	Swiss mice	ID_{50} 1.7 mg.kg^{-1}	Lima et al. 2012
P. aduncum	Larvicidal	*Aedes aegypti*	LD_{50} 0.009 mL e 46 μg/mL	Oliveira et al. 2013, Souza et al. 2020
	Antibacterial	*Staphylococcus epidermidis, S. aureus, Aeromonas hydrophila*	MIC 250 and 500 μg.mL^{-1}	Brazão et al. 2014, Majolo et al. 2019
	Antifungal	*Crinipelis perniciosa*	MIC 0.6 ppm	Almeida et al. 2009
	Biopesticide	*Solenopsis saevissima*	LC_{50} 58.4 mg.mL^{-1}	Souto et al. 2012
	Anti-inflammatory	Wister rats	ID_{50} mg.kg^{-1}	Parise-Filho et al. 2011
	Antifungal	*Colletotrichum musae*	MIC 50 mg.mL^{-1}	Bastos and Albuquerque 2004
	Antifungal	*Sclerotinia sclerotiorum*	MIC 50 μL	Valadares et al. 2018
	Insecticide	*Euschistus heros*	LD_{50} 36.23 mg per insect	Cossolin et al. 2019
P. anonifolium	Antioxidant	–	DPPH 13.0 ± 2.3	Silva et al. 2014
	Antifungal	*Cladosporium cladosporioides, C. sphareospermum*	LD_{50} 10.00 μg	Silva et al. 2014
P. arboretum	Larvicidal	*Aedes aegypti*	LC_{50} 55 μg/mL	Souza et al. 2020
P. enckea	Antifungal	*Crinipellis perniciosa, Phytophthora palmivora, P. capsici*	MIC 1.0 μL.mL^{-1}	Silva and Bastos 2007

...Table 3.2 contd.

...Table 3.2 contd.

Specie	Biological activity	Animal and microorganism	Dose, concentration, and MIC	References
P. marginatum	Repellent	*Tribolium castaneum*	0.01 $\mu g.mL^{-1}$	Jaramillo et al. 2015
	Insecticide	*Spodoptera littoralis*	167 $mg.mL^{-1}$	Souto et al. 2012
	Antibacterial	*Aeromonas hydrophila*	MIC 234.37 to 30,000 $\mu g\ mL^{-1}$	Majolo et al. 2019
	Antifungal	*Cladosporium cladosporioides C. sphareospermum* *Colossoma macropomum*	MIC 1.0 $\mu L.mL^{-1}$	Silva and Bastos 2007, Alves et al. 2021
	Antifungal	*Fusarium oxysporum*	MIC 10 µL	Santos et al. 2011
	Antiparasitic	*Neoechinorhynchus buttnerae*	EC_{50} 1.2 $mg.mL^{-1}$	Santos et al. 2018
	Insecticide	*Zabrotes subfasciatus*	DL_{50} 1.17 µL/L	Ayres et al. 2021
	Ovicidal	*Oebalus insularis*	LC_{50} 1.9 µg/mL	Zachrisson et al. 2019
P. hispidum *P. hispidinervum*	Antioxidant	–	DPPH 26.4 ± 4.2	Silva et al. 2014
	Antifungal	*Cladosporium cladosporioides C. sphareospermum*	DL_{50} 1.00 µg	Silva et al. 2014
	Insecticide	*Spodoptera littoralis*	EC_{50} 48.0 $\mu g.cm^2$	Colorado et al. 2019
	Phytotoxic	*Lolium perenne, Lactuca sativa*	MIC 0.2 and 0.1 mg/mL	Colorado et al. 2019
	Antiparasitic	*Neoechinorhynchus buttnerae*	EC_{50} = 1.7 $mg.mL^{-1}$	Santos et al. 2018
	Antifungal	*Bipolaris sorokiniana, Fusarium oxysporum* e *Colletotrichum gloeosporioides* *Colossoma macropomum*	MIC 100 $\mu g.mL^{-1}$	Zacaroni et al. 2009, Alves et al. 2021
	Antibacterial	*Aeromonas hydrophila*	MIC 234.37 to 30,000 $\mu g.mL^{-1}$.	Majolo et al. 2019
	Larvicidal	*Tanaecium octurnum*	DL_{50} 0.025 µL	Fazolin et al. 2007
	Insecticide	*Spodoptera frugiperda*	CL_{50} 16.2 $mg.mL^{-1}$	Lima et al. 2009
	Antiparasitic	*Neoechinorhynchus buttnerae*	EC_{50} = 0.5 $mg.mL^{-1}$	Santos et al. 2018
P. hostmanianum	Larvicidal	*Aedes aegypti*	LC_{50} 36 µg/mL	Souza et al. 2020
P. gaudichaudianum	Larvicidal	*Aedes aegypti*	LC_{50} 121 µg/mL	Souza et al. 2020
	Antibacterial	*S. aureus*		Silva et al. 2019a

Specie	Biological activity	Animal and microorganism	Dose, concentration, and MIC	References
P. humaytanum	Larvicidal	*Aedes aegypti*	LC_{50} 156 µg/mL	Souza et al. 2020
P. divaricatum	Antifungal	*Fusarium solani*	MIC 0.75 $mg.mL^{-1}$	Meireles 2014
	Insecticide	*Solenopsis saevissima*	LC_{50} 612.5 $mg.mL^{-1}$	Souto et al. 2012
	Antifungal	*Colossoma macropomum*	MIC 1.0 µL/mL	Silva et al. 2014
	Antioxidant	–	DPPH 34.69 ± 1.38	Oliveira et al. 2019
	?Anesthetic	*Colossoma macropomum*	40 $\mu L.L^{-1}$	Vilhena et al. 2019
	Insecticide	Rhopalosiphum padi	EC_{50} 37,2 $\mu g.cm^2$	Colorado et al. 2019
	Phytotoxic	*Lolium perenne,* *Lactuca sativa*	MIC 0.05 and 0.2 mg/mL	Colorado et al. 2019
	Antioxidant	–	DPPH 14.1 ± 0,9	Corpes et al. 2019
P. dilatatum	Antifungal	*Crinipellis perniciosa, Phytophthora capsica*	MIC 1.0 $\mu L.mL^{-1}$	Silva and Bastos 2007
	Insecticide	*Rhopalosiphum padi,* *Myzus persicae*	EC_{50} 8,4 and 35,4 $\mu g.cm^2$	Colorado et al. 2019
P. callosum	Antifungal	*Crinipellis perniciosa, Phytophthora palmivora, P. capsica* *Colossoma macropomum*	MIC 0.75 and 1.0 $\mu L.mL^{-1}$	Silva and Bastos 2007, Alves et al. 2021
	Antibacterial	*Aeromonas hydrophila*	MIC 234.37 to 30,000 µg mL^{-1}.	Majolo et al. 2019
	Insecticide	*Solenopsis saevissima*	LC_{50} 690 $mg.mL^{-1}$	Souto et al. 2012
	Antiparasitic	*Neoechinorhynchus buttnerae*	EC_{50} = 2.1 $mg.mL^{-1}$	Santos et al. 2018
	Insecticide	*Zabrotes subfasciatus*	DL_{50} 0.78 µL/L	Ayres et al. 2021
P. sanctifelicis	Insecticide	*Spodoptera littoralis*	EC_{50} 4,5 $\mu g.cm^2$	Colorado et al. 2019
	Phytotoxic	*Lolium perenne,* *Lactuca sativa*	MIC 0.2 mg/mL	Colorado et al. 2019
P. permucronatum	Larvicidal	*Aedes aegypti*	LC_{50} 36 µg/mL	Souza et al. 2020

...Table 3.2 contd.

...Table 3.2 contd.

Specie	Biological activity	Animal and microorganism	Dose, concentration, and MIC	References
P. carpunya	Antifungal	*Candida albicans, C. glabrata*	MIC 781 and 1562 μg/mL	Ballesteros et al. 2019
	Antioxidant	–	IC_{50} 159.8 μg/mL	Ballesteros et al. 2019
P. jacquemontianum	Ovicidal	*Oebalus insularis*	IC_{50} 1.1 μg/mL	Zachrisson et al. 2019
P. capitarianum	Larvicidal	*Aedes aegypti* and *A. albopictus*	LC_{50} 87.6 and 76.1 μg/mL	França et al. 2021
	Insecticide	*A. aegypti* and *A. albopictus*	LC_{50} 126.2 and 124.5 μg/mL	França et al. 2021
	Anti-acetylcholinesterase (AChE)	Enzyme AChE	IC_{50} 14.2 μg/mL	França et al. 2021
Peperomia serpens	Anti-inflammatory and antinociceptive	Swiss mice	ED_{50} 188.8 $mg.kg^{-1}$	Pinheiro et al. 2011
Piper lhotzkyanum Kunth	Antibacterial	*Mycobacterium tuberculosis*	MIC of 76.51 μg/mL and 128 μg/mL for IFr and IFl	Oliveira et al. 2021

In this study, the antimicrobial activity depended on the size of the particles. The essential oil of *P. nigrum* showed potentiated antibacterial activity when inserted in nano emulsions.

The mechanism of anti-inflammatory activity of pure *P. cubeba* oil and a nano emulsion drug delivery system was evaluated by (Shakeel et al. 2015). The essential oil nano emulsions were developed by an aqueous phase titration method. The pure dose of the essential oil was 400 $mg.kg^{-1}$ of body weight, while the nano emulsion was administered in two different doses, 40 $mg.kg^{-1}$ and 80 $mg.kg^{-1}$ of body weight. The results of the nano emulsion were superior to pure oil, even at significantly lower doses. These results indicate the potential of the nano emulsion developed to improve the therapeutic efficacy of the essential oil of *P. cubeba*.

Lucca et al. (2017) developed nano emulsions of copaiba oil rich in β-caryophyllene, that has a potent anti-inflammatory activity. This reinforces the use of this substance which is also present in *Piper* species of the Amazon. Positively charged Nano emulsions (PCN), with cetyltrimethylammonium bromide, were compared to a Negatively Charged Nan oemulsion (NCN) for *in vitro* skin permeation and *in vivo* antioedematogenic studies. The results showed that cutaneous permeation with PCN increased the retention of caryophyllene in the epidermis three times. The tests *in vivo* on a mouse ear edema, NCN and PCN promoted an inhibition of edema (33%). In the rat paw edema, both nano emulsions had an antiedematogenic effect (edema inhibition above 60%), the nano emulsification process led to an increased effect on the oil. In the following study, the nano emulsions (Lucca et al. 2018) were incorporated in hydrogels prepared with Carbopol®, hydroxyethylcellulose or chitosan. The two former hydrogels showed good stability, droplet size and polydispersity index. In the skin permeation test, it showed high retention in the epidermis followed by less retention in the dermis. In addition, the anti-inflammatory effect was observed with 69 and 67% inhibitions in the mouse ear edema and 32 and 72% in the rat paw edema. Histological sections showed a decrease in inflammatory factors, such as hyperplasia of the dermis and epidermis and inflammatory cells infiltration.

One of the substances found in species of *Piper* from Amazonia is monoterpene linalool. Its insertion in pharmaceutical technology can improve the low aqueous solubility, so an efficient channel is needed to improve its administration and bioavailability in physiological media. Rodenak-Kladniew et al. (2017) studied the encapsulation of linalool using Solid Lipid Nanoparticles (SLN) for *in vitro* testing in two cell lines: hepatocarcinoma and lung adenocarcinoma. The SLN were prepared with myristyl myristate and cetyl palmitate by the sonication method with Pluronic F68 as the surfactant. The NLS showed a spherical shape and diameters in the 90–130 nm range. The percentages of SLN encapsulation were greater than 80% for all formulations and exhibited controlled release profiles *in vitro* for at least 72 hr. The formulations developed, and especially SLN with linalool and myristyl myristate with a concentration of 2.0 mM for 48 hr, showed stronger anticancer activity than free linalool, inhibiting cell growth by 76%. Fluorescence microscopy showed an increase in the capacity of nanoparticles to deliver molecules intracellularly.

To reduce volatilization and improve linalool bioavailability, nanostructured lipid transporters were prepared using the high-pressure homogenization method

(Shi et al. 2016). The rats were treated with 300 $mg.kg^{-1}$ by oral gavage for pharmacokinetic studies. The encapsulation efficiency was 79.6% and all pharmacokinetic and bioavailability parameters of the pharmaceutical formulation were improved compared with pure linalool.

Linalool nano emulsions was also studied for antibacterial activity (Prakash et al. 2019) for use in food systems. With the addition of Tween 80 and water in ultrasound, emulsification systems were developed to overcome the linalool's low water solubility as well as its high volatility and low stability. The nano emulsions were stable with an average droplet diameter of 10.9 ± 0.1 nm, which were obtained at 1 : 3 (*v/v*) ratio of linalool and Tween 80. Increased antibacterial activity against *Salmonella typhimurium* was observed, attributed to an increased ability to permeate the cell membrane. The Minimum Inhibitory Concentration (MIC) and Minimum Bactericidal Concentration (MBC) of linalool for *S. typhimurium* was 1.25 (*v/v*%), whereas in the nano emulsion MIC and MBC was observed in lower concentrations (0.625 *v/v*%). Linalool nano emulsions were also effective in reducing *S. typhimurium* biofilms on the surface of a cut pineapple. Compared with pure linalool, nano emulsions with linalool showed 11.5% greater antibiofilm activity.

Cyclodextrin inclusion complexes with linalool were also studied aiming to the increase of the antimicrobial activity (Aytac et al. 2017). The complex presented rapid dissolution and release of linalool from the complex, inhibited the growth of Gram-negative (*E. coli*) and Gram-positive (*S. aureus*) bacteria. The antibacterial activity of HPβCD/linalool, MβCD/linalool-and HPγCD/linalool was 84 ± 0.1%, 93 ± 0.1% and 95 ± 0.2% against *E. coli* and 70 ± 0.3%, 79 ± 0.2% and 88 ± 0.5% against *S. aureus*, respectively. Briefly, the characteristics of linalool have been preserved and provided high load capacity, a longer shelf life and stronger antibacterial activity than pure linalool.

Linalool in pharmaceutical technology was also studied by (Jabir et al. 2019). Inserted into gold nanoparticles conjugated with a complex with CALNN peptide, its cytotoxicity against breast cancer cell lines was investigated. There was low cytotoxicity in the breast cancer cell line after 48 hr of treatment with linalool at a concentration of 10 $\mu g.mL^{-1}$. The formulation presented antiproliferative and cytotoxic properties at a concentration of 10 $\mu g.mL^{-1}$, being able to kill more than 70% of the cells. The apoptogenic property of the formulation was also investigated. The morphological changes of the breast cancer cell line, analyzed in an inverted phase contrast microscope, showed typical apoptotic characteristics, such as membrane loss.

Nano emulsions containing volatile oils are an important alternative for the control of the *Aedes aegypti* mosquito. The volatile oil from the leaves of *Piper alatipetiolatum*, containing the main active ishwarone (78.6%), had lower ovicidal activity (19 to 100%), larvicidal activity (LC_{50} 33.74 ppm), and pupicidal activity (LC_{50} 65.06 ppm) compared to nano emulsion, which had higher ovicidal activity (47.7 to 100%), larvicidal activity (LC_{50} 6.37 ppm) and pupicidal activity (LC_{50} 9.33 ppm) against *A. aegypti*. These results showed that the nano emulsion increased the bioactive power of the essential oil, interrupting the development of immature forms of *A. aegypti* (Oliveira et al. 2020).

The encapsulated essential oils of *P. aduncum* and *P. hispidinervum* in gelatin nanoparticles were evaluated against *Aedes aegypti* larvae, *Tetranychus urticae* Koch mite and aphid *Cerataphis lataniae* Boisd in studies by (Silva et al. 2019b). The nanoparticles containing *P. aduncum* essential oil showed 100% mortality against larvae and mites for up to 24 hr. About 90% of the mites were killed within 24 hr and total mortality was achieved within 48 hr. Nanoparticles containing the essential oil of *P. hispidinervum* showed 100% mortality against larvae, mites, and aphids for up to 24 hr, being an alternative for combating and controlling the tested pests.

These studies show the importance of inserting essential oils from Amazonian Piperaceae in submicrometric formulations in order to increase their bioefficacy. The biological activities described for *Piper* species emphasize the importance of further studies on the insertion of these bioproducts in pharmaceutical formulations.

Conclusion

Essential oils and their actives from Piperaceae species have important bioactivities for therapeutic and industrial use. Research emphasizes the description of the chemical composition and biological activity of essential oils from the Amazonian Piperaceae species, which reinforces the use of substances from these aromatic oils in pharmaceutical technologies, such as nanostructured and cyclodextrin systems, for the preparation of pharmaceuticals and cosmetics to be more effective, since their bioproducts have great bioactive potential. The application of nanotechnology to the actives of essential oils from Amazonian Piperaceae allow a greater pharmacological and cosmetic exploration of these natural sources, increasing their economic potential, greater knowledge of Amazonian biodiversity and elaboration of new high efficiency bioproducts.

Acknowledgements

The authors acknowledge for the support Brazilian National Council for Scientific and Technological Development (CNPq), Brazilian Coordination for the Improvement of Higher Education Personnel (CAPES), Foundation for Research Support of the State of Amazonas (FAPEAM).

References

Alencar, R.D., R. Lima, C.R. Corrêa, O.R. Gottlieb, M.C. Marx, M. Silva, et al. 1971. Óleos essenciais de plantas brasileiras. Acta Amaz. 1: 41–43.

Almeida, R.R., R.N. Souto, C.N. Bastos, M.H. Silva and J.G. Maia. 2009. Chemical variation in *Piper aduncum* and biological properties of its dillapiole rich essential oil. Chem. Biodivers. 6: 1427–1434.

Alves, N.S.F., W.N. Setzer and J.K.R. Silva. 2019. The chemistry and biological activities of *Peperomia pellucida* (Piperaceae): A critical review. J. Ethnopharmacol. 232: 90–102.

Alves, C.M.G., J.N. Nogueira,, J.G.R. Luz, F.C.M. Chaves and M.T. Dias. 2021. Essential oil of *Piper callosum*, *Piper hispidum* and *Piper marginatum* (Piperaceae) possesses *in vitro* efficacy against monogeneans of *Colossoma macropomum* (tambaqui). Aquacult. Res. 1: 1–10.

Anderson, R.R., N. Girola, C.R. Figueiredo, V.S. Londero and J.H.G. Lago. 2018. Avaliação da variação circadiana e da atividade citotóxica *in vitro* de compostos voláteis das folhas de *Piper regnellii* (Miq) C. DC. var. *regnellii* (C. DC.) Yunck (Piperaceae). Nat. Prod. Res. 32: 859–862.

Andrade, E.H.A, C.N. Alves, E.F. Guimarães, L.M.M. Carreira and J.G.S. Maia. 2011. Variability in essential oil composition of *Piper dilatatum* LC Rich. Biochem. System. Ecol. 39: 669–675.

Araújo, C.A., C.A.G. Camara, M.M. Moraes, G.J.N. Vasconcelos, M.R.S. Pereira and C.E. Zartman. 2018. First record of the chemical composition of essential oil of *Piper bellidifolium*, *Piper durilignum*, *Piper acutilimbum* and *Piper consanguineum* from the Brazilian Amazon Forest. Acta Amaz. 48: 330–337.

Arevalo, M.J.V., S.M. Cruz, M.A. Apel, A.T. Henriques and A. Caceres. 2019. Essential oil of *Piper oradendron* from the pacific slope of Guatemala. Nat. Prod. Comm. 14: 79–81.

Autran, E.S., I.A. Neves, C.S.B. Silva, G.K.N. Santos, C.A.G. Câmara and D.M.A.F. Navarro. 2009. Chemical composition, oviposition deterrent and larvicidal activities against *Aedes aegypti* of essential oils from *Piper marginatum* Jacq. (Piperaceae). Biores. Technol. 100: 2284–2288.

Ayres, V.F.S., M.R. Oliveira, E.L.L. Baldin, G.M. Corrêa, A.C. Guimarães and R. Takeara. 2021. Chemical composition and insecticidal activity of the essential oils of *Piper marginatum*, *Piper callosum* and *Vitex agnus-castus*. Ann. Braz. Acad. Sci. 93: 1–11.

Aytac, Z., I. Z. Yildiz, F. Kayaci-Senirmak, T. Tekinay and T. Uyar. 2017. Electrospinning of cyclodextrin/linalool-inclusion complex nanofibers: Fast-dissolving nanofibrous web with prolonged release and antibacterial activity. Food Chem. 231: 192–201.

Ballesteros, J.L., M. Tacchini, A. Spagnoletti, A. Grandini, G. Paganetto, L.M. Neri, et al. 2019. Rediscovering medicinal Amazonian aromatic plants: *Piper carpunya* (Piperaceae) essential oil as paradigmatic study. Evid. Based Complement. Alternat. Med. 2019: 1–10.

Barbosa, Q.P., C.A. Câmara, C.S. Ramos, D.C. Nascimento, J.V. Lima-Filho and E.F. Guimarães. 2012. Chemical composition, circadian rhythm and antibacterial activity of essential oils of *Piper divaricatum*: a new source of safrole. Quím. Nova. 35: 1806–1808.

Barreiro, E.J. and C.A. Fraga. 1999. A utilização do safrol, principal componente químico do óleo de sassafráz, na síntese de substâncias bioativas na cascata do ácido araquidônico: anti-inflamatórios, analgésicos e antitrombóticos. Quím. Nova. 22: 744–759.

Basak, S. and P. Guha. 2017. Betel leaf (*Piper betle* L.) essential oil microemulsion: Characterization and antifungal activity on growth, and apparent lag time of *Aspergillus flavus* in tomato paste. Lebensmittel-Wissenschaft Technol. 75: 616–623.

Bastos, C.N. and P.S.B. Albuquerque. 2004. Efeito do óleo de *Piper aduncum* no controle em pós-colheita de *Colletotricum musae* em banana. Fitopatol. Bras. 29: 555–557.

Bizzo, H.R., A.M.C. Hovell and C.M. Rezende. 2009. Óleos essenciais no Brasil: aspectos gerais, desenvolvimento e perspectivas. Quím. Nova. 32: 588–594.

Bilia A.R., C. Guccione, B. Isacchi, C. Righeschi, F. Firenzuoli and C.M. Bergonzi. 2014. Essential oils loaded in nanosystems: a developing strategy for a successful therapeutic approach. Evid Based Complement Alternat Medicine. 2014: 651593.

Branquinho, S.L., A.J. Santosa, L.A.C. Cardoso, S.J. Mota, U. Lanza-Junior, L.A.C. Kassuya, et al. 2017. Anti-inflammatory and toxicological evaluation of essential oil from *Piper glabratum* leaves. J. Ethnopharmacol. 198: 372–378.

Brazão, M.A.B., F.V. Brazão, J.G.S. Maia and M.C. Monteiro. 2014. Antibacterial activity of the *Piper aduncum* oil and dillapiole, its main constituent, against multidrug-resistant strains. Bol. Latinoam. Caribe Pl. Med. Arom. 13: 517–526.

Burger, W.C. 1971. Piperaceae. pp. 5–218. *In:* Burger, W.C., (ed.). Fieldiana Botany: Flora Costaricensis. Chicago: Field Museum of Natural History.

Cáceres, A., S.M. Cruza, J.V. Martínez-Arevalob, A.T. Henriquesc and M.A. Apelc. 2019. Composition of essential oil from *Piper jacquemontianum* from eight provenances of Guatemala. Nat. Prod. Comm. 14: 75–78.

Carneiro, S.B., F.I.C. Duarte, L. Heimfarth, S.J.S. Quintans, J.L. Quintans-Júnior, V.F. Veiga Júnior, et al. 2019. Cyclodextrin–Drug Inclusion Complexes: *In Vivo* and *In Vitro* Approaches. Int. J. Mol. Sci. 20: 642.

Cheirsilp, B. and J. Rakmai. 2017. Inclusion complex formation of cyclodextrin with its guest and their applications. Biol. Eng. Med. 2: 1–6.

Colorado, J.B.E., N. Benitez and P.A.C González. 2019. Volatile composition and biocidal (antifeedant and phytotoxic) activity of the essential oils of four Piperaceae from Choco-Colombia. Ind. Crops Prod. 138: 111463.

Corpes, R., L. Monteiro, E.H. Andrade, J.G. Maia, I. Menezes, W.N. Setzer, et al. 2019. Comparison of volatile profile and antioxidant activity of *Piper divaricatum* G. Meyer (Piperaceae) using cuttings and cell tissue. J. Braz. Chem. Soc. 30: 2291–2298.

Corral, T.C.A. 2014. Uso do óleo essencial de *Piper aduncum* na ração para o controle de nematóides de pirarucu (*Arapaima gigas*, SCHINZ, 1822). pp. 67. Dissertação de mestrado (Aquicultura) – Programa de Pós-Graduação em Aquicultura. Universidade Nilton Lins. Manaus, Amazonas.

Cossolin, J.F.S., M.J.B. Pereira, L.C. Martínez, L.M. Turchen, M. Fiaz, H. Bozdoğan, et al. 2019. Cytotoxicity of *Piper aduncum* (Piperaceae) essential oil in brown stink bug *Euschistus heros* (Heteroptera: Pentatomidae). Ecotoxicol. 28: 763–770.

Cremasco, M.A. and N.P. Braga. 2010. Tecnologia de Processamento do óleo essencial de pimentalonga *(Piper hispidinervum* C. DC.) para a produção de piperonal. pp. 1–11. *In:* Associação Brasileira de Engenharia Química (ed.). Anais do XVIII Congresso Brasileiro de Engenharia Química. Foz do Iguaçu: Associação Brasileira de Engenharia Química.

Cronquist, A. 1981. An integrated system of classification of flowering plants. Reissue. New York: Columbia University Press.

Cysne, J.B., K.M. Canuto, O.D.L. Pessoa, E.P. Nunes and E.R. Silveira. 2005. Leaf essential oils of four *Piper* species from the state of Ceará—northeast of Brazil. J. Braz. Chem. Soc. 16: 1378–1381.

Damasceno, B.P.G.L., J.A. Silva, E.E. Oliveira, W.L.L. Silveira, I.B. Araújo, A.G. Oliveira, et al. 2011. Microemulsão: um promissor carreador para 63 moléculas insolúveis. Rev. Ci. Farm. Bás. Apl. 32: 9–18.

Durofil, A., M. Radice, J. Blanco-Salas and T. Ruiz-Téllez. 2021. *Piper aduncum* essential oil: a promising insecticide, acaricide and antiparasitic. A review. Parasite. 28: 42.

Elisabetsky, E. 1991. Sociopolitical, economical and ethical issues in medicinal plant research. J. Ethnopharmacol. 32: 235–239.

Evans, P.H., W.S. Bowers and E.J. Funk. 1984. Identification of fungicidal and nematocidal components in the leaves of *Piper betle* (Piperaceae). J. Agr. Food Chem. 32: 1254–1256.

Facundo, V.A., A.R. Pollli, R.V. Rodrigues, J.S.L.T. Militão, R.G. Stabelli and C.T. Cardoso. 2008. Constituintes químicos fixos e voláteis dos talos e frutos de *Piper tuberculatum* Jacq. e das raízes de *P. hispidum* H.B.K. Acta Amaz. 38: 733–742.

Fanun, M. 2012. Microemulsions as delivery systems. Cur. Op. Coll. Interf. Sci. 17: 306–313.

Fazolin, M., J.L.V. Estrela, V. Catani, M.R. Alécio and M.S. Lima. 2007. Propriedade inseticida dos óleos essenciais de *Piper hispidinervum* C. D.C.; *Piper aduncum* L. *e Tanaecium nocturnum* (Barb. Rodr.) Bur. & K. Shum sobre *Tenebrio molitor* L. Ci. Agrotecnol. Lavras. 31: 113–120.

Fernandes, E.S., G.F. Passos, R. Medeiros, F.M. da Cunha, J. Ferreira, M.M. Campos, et al. 2007. Anti-inflammatory effects of compounds α-humulene and β-caryophyllene isolated from the essential oil of *Cordia verbenacea*. Eur. J. Pharmacol. 569: 228–236.

França, L.P., A.C.F. Amaral, A.S. Ramos, J.L. Pereira, A.C.B. Maria, K.M.T. Oliveira, et al. 2021. *Piper capitarianum* essential oil: a promising insecticidal agent for the management of *Aedes aegypti* and *Aedes albopictus*. Environ. Sci. Pollut. Res. 28: 9760–9776.

Frodin, D.G. 2004. History and concepts of big plant genera. Taxon. 53: 753–776.

Gartner, B.L. 1989. Breakage and regrowth of *Piper* species in rain forest understory. Biotropica. 21: 303–307.

Gallardo, K.C., J. Olivero-Verbel, N. Pino-Benítez and E.E. Stashenko. 2014. Chemical composition and bioactivity of *Piper auritum* and *P. multiplinervium* essential oils against the red flour beetle, *Tribolium castaneum* (Herbst). Bol. Latinoam. Caribe Pl. Med. Arom. 13: 10–19.

Gilardoni, G., Y. Matute and J. Ramírez. 2020. Chemical and enantioselective analysis of the leaf essential oil from *Piper coruscans* Kunth (Piperaceae), a costal and amazonian native species of Ecuador. Plants. 24: 791.

Gogosz, A.M., M.R.T. Boeger, R.R.B. Negrelle and C. Bergo. 2012. Anatomia foliar comparativa de nove espécies do gênero *Piper* (Piperaceae). Rodriguésia. 63: 405–417.

Grimshaw, J. and C. Hua. 1994. The conversion of isosafrole to piperonal and anethole to anisadehyde: electrochemical active manganese dioxide. Electrochim. Acta. 39: 497–499.

Guimarães, E.F. and M.C. Valente. 2001. *Piperaceae–Piper*. pp. 4–91. *In:* A. Reis (ed.). Flora ilustrada catarinense. Itajaí: Herbário Barbosa Rodrigues.

Guimarães, E.F. and L.C.S. Giordano. 2004. Piperaceae do nordeste brasileiro I: estado do Ceará. Rodriguésia. 55: 21–46.

Gupta, M.P., T.D. Arias, N.H. Williams, R. Bos and D.H.E. Tatije. 1985. Safrole, the main component of the essential oil from *Piper aurztum* of Panama. J. Nat. Prod., 48: 330–343.

Jabir, S.M., A.A. Taha, J.Z. Taqi, M.A. Al-Shammari and S.A. Salman. 2019. Novel of nano delivery system for Linalool loaded on gold nanoparticles conjugated with CALNN peptide for application in drug uptake and induction of cell death on breast cancer cell line. Mat. Sci. Eng. 94: 949–964.

Jaramillo, C.J., T.E. Julio, R.A. Duarte, C.L.F. Gonzalez and J. Torres. 2015. Estudio comparativo de la composición volátil y las actividades biológicas del aceite essential de *Piper marginatum* Jacq. Colombiano. Bol. Latinoam. Pl. Med. Arom. 14: 343–354.

Jirovetz, L., G. Buchbauer, M.B. Nagassoum and M. Geissler. 2002. Aroma compound analysis of *Piper nigrum* and *Piper guineese* essential oils from Cameroon using solid phase microextraction gas chromatography, solid phase micoextraction gas chromatography mass spectrometry and olfactometry. J. Chromat. A. 976: 265–275.

Jímenez, M., A. Jazmín, A. Domínguez, A. Luz, B. Pascual-Pineda, A. Ebner Azuara, et al. 2017. Elaboration and characterization of O/W cinnamon (*Cinnamomum zeylanicum*) and black pepper (*Piper nigrum*) emulsions. Food Hydrocoll. 2017: 1–9.

Kyriakoudi, A., E. Spanidi, I. Mourtzinos and K. Gardikis. 2021. Innovative delivery systems loaded with plant bioactive ingredients: Formulation approaches and applications. Plants (Basel). 10: 1238.

Klauke, A.L., I. Racza, B. Pradiera, A. Markerta, M.A. Zimmera, J. Gertschb, et al. 2014. The cannabinoid CB2 receptor-selective phytocannabinoid beta-caryophyllene exerts analgesic effects in mouse models of inflammatory and neuropathic pain. Eur. Neuropsychopharmacol. 24: 608–620.

Krinski, D., L.A. Foerster and C. Deschamps. 2018. First phytochemical description of essential oils from *Piper cachimboense* (Piperales, Piperaceae). Acta Amaz. 48: 70–74.

Lee, S.E., B.S. Park, M.K. Kim, W.S. Choi, H.T. Kim, K.Y. Cho, et al. 2001. Fungicidal activity of pipernonaline, a piperidine alkaloid derived from long pepper, *Piper longum* L., against phytopathogenic fungi. Crop Protection. 20: 523–528.

Lima, R.K., M.G. Cardoso, J.C. Moraes, B.A. Melo, V.G. Rodrigues and P.L. Guimarães. 2009. Atividade inseticida do óleo essencial de pimenta longa (*Piper hispidinervum* C. D.C.) sobre lagarta-do-cartucho do milho *Spodoptera frugiperda* (J.E. Smith, 1797) (Lepidoptera: Noctuidae). Acta Amaz. 39: 377–382.

Lima, D.K., L.J. Ballico, L.F. Rocha, H.P. Gonçalves, L.M. Souza, M. Iacomini, et al. 2012. Evaluation of the antinociceptive, anti-inflammatory and gastric antiulcer activities of the essential oil from *Piper aleyreanum* C. D.C. in rodents. J. Ethnopharmacol. 142: 274–282.

Lima, P.S.S., M.A. Lucchese, G.H. Araújo-Filho, P.P. Menezes, A.A.S. Araújo, J.L. Quintans-Júnior et al. 2016. Inclusion of terpenes in cyclodextrins: Preparation, characterization and pharmacological approaches. Carbohydr. Pol. 151: 965–987.

Lucca, G.L., P.S. Matos, B.C. Mattos, F.H. Teixeira, P.R. Limberger, V.F. Veiga-Júnior, et al. 2017. Nanoemulsification Potentiates *In Vivo* Antiedematogenic Effect of Copaiba Oil. J. Biomed. Nanotechnol. 13: 583–590.

Lucca, G.L., P.S. Matos, T. Kreutz, F.H. Teixeira, P.R. Limberger, V.F. Veiga-Júnior, et al. 2018. Anti-inflammatory Effect from a Hydrogel Containing Nanoemulsified Copaiba oil (*Copaifera multijuga* Hayne). AAPS PharmSciTech. 19: 522–530.

Majolo, C., P.C. Monteiro, A.V.P. Nascimento, F.C.M. Chaves, P.E. Gama, H.R. Bizzo, et al. 2019. Óleos essenciais de cinco espécies brasileiras de *Piper* como antimicrobianos contra cepas de *Aeromonas hydrophila*. J. Essent. Oil Bearing Plants. 22: 746–761.

Maia, J.G.S., M.L. Silva, A.I.R. Luiz, M.G.B. Zoghbi and L.S. Ramos. 1987. Espécies de *Piper* da Amazônia ricas em safrol. Quím, Nova. 10: 200–204.

Maia, J.G.S., M.G.B. Zohhbi, E.H.A. Andrade, A.S. Santos, M.H.L. Silva and C.N. Bastos. 1998. Constituents of the essential oil of *Piper aduncum* L. growing wild in the Amazon region. Flav. Frag. J. 13: 269–272.

Marques, A.M., A.C.C. Peixoto, R.C. Paula, M.F.A. Nascimento, L.F. Soares, L.S.M. Velozo, et al. 2015. Phytochemical investigation of anti-plasmodial metabolites from Brazilian native *Piper* species. J. Ess. Oil-Bearing Plants. 18: 74–81.

Matos, S.P., F.H., À.A.N. Lima, V.F. Veiga-Junior and L.S. Koester. 2019. Essential oils and isolated terpenes in nanosystems designed for topical administration: A review. Biomolecules. 9: 138.

Meireles, N.S. 2014. Influência dos metabólitos secundários de *Piper divaricatum* da região amazônica no controle do *Fusarium solani* f. sp. piperis causador da fusariose em pimenta do reino. pp. 183. Federal University of Pará (Masters dissertation). Belém, Pará, Brazil.

Mesquita, J.M.O., C. Cavaleiro, A.P. Cunha, J.A. Lombardi and A.B. Oliveira. 2005. Estudo comparativo dos óleos voláteis de algumas espécies de Piperaceae. Braz. J. Pharmacog. 15: 6–12.

Monzote, L., M. García, A.M. Montalvo, R. Scull and M. Miranda. 2010. Chemistry, cytotoxicity and antileishmanial activity of the essential oil from *Piper auritum*. Mem. Inst. Oswaldo Cruz. 105: 168–173.

Moraes, S.M., V.A. Facundo, L.M. Bertini, E.S.B. Cavalcanti, J.F.A. Júnior, A.S. Ferreira, et al. 2007. Chemical composition and larvicidal activity of essential oils from *Piper* species. Biochem. Syst. Ecol. 35: 670–675.

Mundina, M., R. Vila, F. Tomi, M.P. Gupta, T. Adzet, J. Casanova, et al. 1998. Leaf essential oils of three panamanian *Piper* species. Phytochemistry. 47: 1277–1282.

Nascimento, A.S., A.E. Araújo, M.J. Silva and S.C. Ramos 2015. Chemical study and antimicrobial activities of *Piper arboreum* (Piperaceae). J. Chilean Chem. Society. 60: 2837–2839.

Navickiene, H.M.D., A.A. Morandim, A.C. Alécio, L.O. Regasini, D.C.B. Bergamo, M. Telascrea, et al. 2006. Composition and antifungal activity of essential oils from *Piper aduncum, Piper arboreum* and *Piper tuberculatum*. Quím. Nova. 29: 467–470.

Negreiros, J.R.S., D.P. Miqueloni and C.B.C. Cartaxo. 2015. Yield of essential oil and safrole content based on fresh and dry biomass of long Pepper in the Brazilian Amazon. Acta Amaz. 45: 75–80.

Noriega, P., A. Guerrini, G. Sacchetti, A. Grandini, E. Ankuash and S. Manfredini. 2019. Chemical composition and biological activity of five essential oils from the ecuadorian amazon rain forest. Molecules. 24: 1–12.

Oliveira, G.L., S.K. Cardoso, C.R. Lara-Júnior, T.M. Vieira, E.F. Guimarães, L.S. Figueiredo, et al. 2013. Chemical study and larvicidal activity against *Aedes aegypti* of essential oil of *Piper aduncum* L. (Piperaceae). An. Acad. Bras. Ci. 4: 1227–1234.

Oliveira, S.M., N.J. Cruze, G.S. Silva, A.W. Costa, B.H.S. Sousa, F.W.F. Bezerra, et al. 2019. Phytochemical profile, antioxidant activity, inhibition of acetylcholinesterase and interaction mechanism of the major components of the *Piper divaricatum* essential oil obtained by supercritical CO_2. J. Supercrit. Fluids. 145: 74–84.

Oliveira, A.C. de, I.S.C. Sá, R.S. Mesquita, B.L. Pereira, L.A. Pocrifka, T.P. de Souza, et al. 2020. Nanoemulsion loaded with volatile oil from *Piper alatipetiolatum* as an alternative agent in the control of *Aedes aegypti*. Revi. Bras. Farmacog. 30: 667–677.

Oliveira, C.C., Y.J. Ramos, G.A. Queiroz, E.F. Guimarães, M.H. Araujo, E.B. Lassounskai, et al. 2021a. Antimycobacterial activity and chemical characterization of the essential oils from reproductive organs of *Piper lhotzkyanum* Kunth (Piperaceae). Rev. Virtual Quím. 13:1196–1202.

Oliveira, M.R., L.A. Silva, R.S. Silva, C.C.B. Queiroz and R. Takeara. 2021b: Chemical composition and biological activities of essential oils of *Piper* species from the Amazon. J. Essent Oil Res. 33: 536–548.

Parise-Filho, R., M. Pastrello, C.E.P. Camerlingo, G.J. Silva, L.A. Agostinho, T. Souza, et al. 2011. The anti-inflammatory activity of dillapiole and some semisynthetic analogues. Pharm. Biol. 49: 1173–1179.

Parmar, V.S., S.C. Jain, K.S. Bisht, R. Jain, P. Taneja, A. Jha, et al. 1997. Phytochemistry of the genus *Piper*. Phytochemistry. 46: 597–673.

Parra, J.E., W.A. Delgado and L.E. Cuca. 2011. Cumanensic acid, a new chromene isolated from *Piper* cf. cumanense Kunth. (Piperaceae). Phytochem. Lett. 4: 280–282.

Pimentel, F.A., M.M.M. Souza, C.P. Sá, W.G. Cabral, M.R. Silva, P.S.N. Pinheiro, et al. 1998. Recomendações básicas para o cultivo da pimenta longa (*Piper hispidinervium*) no Estado do Acre. pp. 1–17 1st edition. Rio Branco: Embrapa-CPAF/AC.

Pinheiro, B.G., A.S. Silva, G.E. Souza, J.G. Figueiredo, F.Q. Cunha, S. Lahlou, et al. 2011. Chemical composition, antinociceptive and anti-inflammatory effects in rodents of the essential oil of *Peperomia serpens* (Sw.) Loud. J. Ethnopharmacol. 8: 479–486.

Pinheiro, J.O.P., T.E. Aragão, S. Santos, F. Juliana, Y. Maria, B. Gomes, et al. 2017. Inclusion complexes of copaiba (*Copaifera multijuga* Hayne) oleoresin and cyclodextrins: Physicochemical characterization and anti-inflammatory activity. Int. J. Mol. Sci. 18: 2388.

Potzernheim, L.C.M., R.H. Bizzo, P.J. Silva and F.R. Vieira. 2012. Chemical characterization of essential oil constituents of four populations of *Piper aduncum* L. from Distrito Federal, Brazil. Biochem. Syst. Ecol. 46: 25–31.

Prakash, A., V. Vadivel, D. Rubini and P. Nithyanand. 2019. Antibacterial and antibiofilm activities of linalool nanoemulsions against *Salmonella typhimurium*. Food Biosci. 28: 57–65.

Ramírez, J., M.D. Andrade, G. Vidari and G. Gilardoni. 2021. Essential oil and major non-volatile secondary metabolites from the leaves of amazonian *Piper subscutatum*. Plants. 10: 1168.

Ramos, Y.J., D.B. Machado, G.A. Queiroz, E.F. Guimarães, A.C.A Defaveri and D.L. Moreira. 2020. Chemical composition of the essential oils of circadian rhythm and of different vegetative parts from *Piper mollicomum* Kunth—A medicinal plant from Brazil. Biochem. Syst. Ecol. 92: 104–116.

Ramos, Y.J., C.D. Oliveira, C.I. Fonseca, G.A.D. Queiroz, E.F. Guimarães, A.C.A.E. Defaveri, et al. 2021. Advanced chemophenetic analysis of essential oil from leaves of *Piper gaudichaudianum* Kunth (Piperaceae) using a new reduction-oxidation index to explore seasonal and circadian rhythms. Plants. 10: 2116.

Rakmai, J., B. Cheirsilp, C.J. Mejuto, A.T. Agrasar and J.S. Gandara. 2017. Physico-chemical characterization and evaluation of bio-efficacies of black pepper essential oil encapsulated in hydroxypropyl-betacyclodextrin. Food Hydrocoll. 65: 157–164.

Raut, J.S. and S.M. Karuppayil. 2014. A status review on the medicinal properties of essential oils. Ind. Crops Prod. 62: 250–264.

Rodenak-Kladniew, B., A.G. Islan, G.M. Bravoa, N. Durán and R.G. Castro. 2017. Design, characterization and *in vitro* evaluation of linalool-loaded solid lipid nanoparticles as potent tool in cancer therapy. Colloids and Surfaces B: Biointerfaces. 154: 123–132.

Rossa, G.E., R.N. Almeida, R.M.F. Vargas, E. Cassel and G. Moyna. 2018. Sequential extraction methods applied to *Piper hispidinervum*: An improvement in the processing of natural products. Can. J. Chem. Eng. 96: 756–762.

Shakeel, F., G.A. Shazly, R. Mohammad, A. Ahmad, M.A. Kalam, N. Ali, et al. 2015. Biological investigation of a supersaturated self-nanoemulsifying drug delivery system of *Piper cubeba* essential oil. RSC Advances. 5: 105–206.

Salehi, B., Z.A. Zakaria, R. Gyawali, S.A. Ibrahim, J. Rajkovic, Z.K. Shinwari, et al. 2019. *Piper* species: A comprehensive review on their phytochemistry, biological activities and applications. Molecules. 24: 071364.

Santos, A.S., A.H.E. Andrade, B.G.M. Zoghbi, R.I.A. Luz and S.G.J. Maia. 1998. Sesquiterpenes of amazonian *Piper* species. Acta Amaz. 28: 127–130.

Santos, M.R.A., R.A. Lima, C.F. Fernandes, A.G. Silva and V.A. Facundo. 2011. Antifungal activity of *Piper marginatum* L. (Piperaceae) essential oil against *Fusarium oxysporum* (Schlecht). Rev. Saúde Pesq. 4: 9–14.

Santos, T.G., R.A. Rebelo, E.M. Dalmarco, A. Guedes, A.L. Gasper, A.B. Cruz, et al. 2012. Composição química e avaliação da atividade antimicrobiana do óleo essencial das folhas de *Piper malacophyllum* (c. Presl.) C. DC. Quím. Nova. 35: 477–481.

Santos, S.P., M.K.L. Souza, L.S.T. Araujo, R.V.J. Medeiros, C.C.S. Nunes, J.B.F. Veiga, et al. 2017. Methyl-β-cyclodextrin inclusion complex with β-caryophyllene: Preparation, characterization, and improvement of pharmacological activities. ACS Omega. 2: 9080–9094.

Santos, W.B., C. Majolo, D. Silva, B. Oliveira, F. Célio, M. Chaves, et al. 2018. Eficácia *in vitro* de óleos essenciais de espécies de Piperaceae no controle do acantocéfalo *Neoechinorhynchus buttnerae*. Rev. Bras. Hig. San. An. 12: 460–469.

Sauter, I.P., G.E. Rossa, A.M. Lucas, S.P. Cibulski, P.M. Roehe, L.A.A. Silva, et al. 2012. Chemical composition and amoebicidal activity of *Piper hispidinervum* (Piperaceae) essential oil. Ind. Crops Prod. 40: 292–295.

Shi, F., Y. Zhao, C.K. Firempong and X. Xu. 2016. Preparation, characterization and pharmacokinetic studies of linalool-loaded nanostructured lipid carriers. Pharmac. Biol. 54: 2320–2328.

Silva, E.M.J. and S.R. Machado. 1999. Estrutura e desenvolvimento dos tricomas secretores em folhas de *Piper regnellii* (Miq.) C. DC. var. regnellii (Piperaceae). Rev. Bras. Bot. 22: 117–124.

Silva, H.M.M.D. and N.C. Bastos. 2007. Atividade antifúngica de óleos essenciais de espécies de *Piper* sobre *Crinipellis perniciosa*, *Phytophthora palmivora* e *Phytophthora capsici*. Fitopatol. Bras. 32: 143–145.

Silva, W.C., J.R.S. Martins, H.E.M. Souza, H. Heinzen, M.V. Cesio, M. Mato, et al. 2009. Toxicity of *Piper aduncum* L. (Piperales: Piperaceae) from the Amazon forest for the cattle tick *Rhipicephalus (Boophilus) microplus* (Acari: Ixodidae). Vet. Parasitol. 164: 267–274.

Silva, J.K., L.C. Pinto, R.M.R. Burbano, R.C. Montenegro, E.F. Guimarães, E.H.A. Andrade, et al. 2014. Essential oils of Amazon *Piper* species and their cytotoxic, antifungal, antioxidant and anti-cholinesterase activities. Ind. Crops Prod. 58: 55–60.

Silva, J.K., R. Trindade, N.S. Alves, P.L. Figueiredo, J.G.S. Maia and W.N. Setzer. 2017. Essential oils from neotropical *Piper* species and their biological activities. Int. J. Mol. Sci. 18: 122–571.

Silva, M.A., G.M. Passarini, N.L. Martinez, V.A. Facundo, B.G.C. Teles and C.C. Kuehn. 2019a. Chemical constituents and bioactivities of essential oils from the genus *Piper* (Piperaceae): A review. South Am. J. Basic Ed. Tech. Technol. 6: 776–817.

Silva, L.S., M.J. Mar, G.S. Azevedo, S.M. Rabelo, A.J. Bezerra, H.P. Campelo, et al. 2019b. Encapsulation of *Piper aduncum* and *Piper hispidinervum* essential oils in gelatin nanoparticles: a possible sustainable control tool of *Aedes aegypti*, *Tetranychus urticae* and *Cerataphis lataniae*. J. Sci. Food Agricul. 99: 685–695.

Silva, A.C.A., E.F. Matias, J.E. Rocha, A.C.J. Araújo, T.S. Freitas, F.F. Campina, et al. 2021. Gas chromatography coupled to mass spectrometry (GC-MS) characterization and evaluation of antibacterial bioactivities of the essential oils from *Piper arboreum* Aubl., *Piper aduncum* L. e *Piper gaudichaudianum* Kunth. Z. Naturforsch. 76: 35–42.

Simões, C.M.O., E.P. Schenkel, G. Gosmann, J.C.P. Mello, L.A. Mentz and P.R. Petrovick. 2007. Farmacognosia: da planta ao medicamento. pp. 103–105. 6th edition. Porto Alegre: Editora da UFRGS.

Simpson, B.B. and M.O. Ogorzaly. 1995. Economic Botany: plants in our world. pp. 78–79. 2nd. edition. New York: McGraw-Hill.

Spear, N.B., R. Awad, P. Maquin, V. Cal, P. Sanchez-Vindas, L. Poveda, et al. 2005. Plant use by the Q'Eqchi' Maya of Belize in ethnopsychiatry and neurological pathology. Econ. Botany. 59: 326–336.

Souto, R.N.P., A.Y. Harada, E.H.A. Andrade and J.G.S. Maia. 2012. Insecticidal activity of *Piper* essential oils from the Amazon against the fire ant *Solenopsis saevissima* (smith) (hymenoptera: formicidae). Neotr. Entomol. 41: 510–517.

Souza, V.C. and H. Lorenzi. 2005. Botânica sistemática: guia ilustrado para identificação das famílias da Angiospermas da flora brasileira, baseado em APGII. p. 56. 2nd. edition. Nova Odessa: Instituto Plantarum. Nova Odessa, SP.

Souza, M.A., L. Silva, M.A.C. Santos, M.J.F. Macêdo, L.J. Lacerda-Neto, H.D.M. Coutinho, et al. 2020. Larvicidal activity of essential oils against *Aedes aegypti* (Diptera: Culicidae). Cur. Pharm. Design. 26: 4092– 4111.

Sotto, A.D., G. Mazzant, I.F. Carbone, P. Hrelia and F. Maffei. 2010. Inhibition by trans-caryophyllene of ethyl methanesulfonate-induced clastogenicity in cultured human lymphocytes. Mut. Res. Gen. Toxicol. Environment. Mutagen. 699: 23–28.

Stevens, P.F. 2017. Angiosperm Phylogeny Website. URL: http: //www.mobot.org/MOBOT/research/APweb/. Accessed on 14/12/2021.

Taylor, D.W. and L.J. Hickey. 1992. Phylogenetic evidence for the herbaceous origin of angiosperms. Plant Syst. Evol. 180: 137–156.

Tripathi, A.K., D.C. Jain and S. Kumar. 1996. Secondary metabolites and their biological and medicinal activities of *Piper* species plants. J. Med. Arom. Plant Sci. 18: 302–321.

Valadares, A.C.F., C.A.F.A. Alves, J.M. Alves, I.P.B. Deus, T.C.L. Santos, H.J. Dias, et al. 2018. Essential oils from *Piper aduncum* inflorescences and leaves: chemical composition and antifungal activity against Sclerotinia sclerotiorum. An. Acad. Bras. Ci. 90: 2691–2699.

Vilhena, S.C., S.A.L. Nascimento, A.H.E. Andrade, R.K.J. Silva, J. Hamoy, F.M. Torres, et al. 2019. Essential oil of *Piper divaricatum* induces a general anaesthesia-like state and loss of skeletal muscle tonus in juvenile tambaqui, *Colossoma macropomum*. Aquaculture. 510: 169–175.

Volpe, H.X.L., M. Fazolin, R.B. Garcia, R.F. Magnani, J.C. Barbosa and M.P. Miranda. 2015. Efficacy of essential oil of *Piper aduncum* against nymphs and adults of *Diaphorina citri*. Soc. Chem. Ind. 72: 1242–1249.

Yuncker, T.G. 1972. The Piperaceae of Brazil I: *Piper*—Group I, II, III, IV. Hoehnea. 2: 19–366.

Zacaroni, L.M., M.G. Cardoso, P.E. Souza, F.A. Pimentel, G. Guimaraes and A.P. Salgado. 2009. Potencial fungitóxico do óleo essencial de *Piper hispidinervum* (pimenta longa) sobre os fungos fitopatogênicos *Bipolaris sorokiniana, Fusarium oxysporum e Colletotrichum gloeosporioides*. Acta Amaz. 39: 193–198.

Zachrisson, B., A. Santana and M. Gupta. 2019. Effects of Essential Oils from Two Species of Piperaceae on Parasitized and Unparasitized Eggs of *Oebalus insularis* (Heteroptera: Pentatomidae) by *Telenomus podisi* (Hymenoptera: Platygastridae). Nat. Prod. Comm. 14: 83–84.

Section 2

Essential Oils in Food and Agriculture

4

Use of Essential Oils in Food Industries—Application as Functional Compounds in Food Matrices and Packaging

Amanda da Silva Ramos[1], *Pedro Henrique Campelo Félix*[2] *and Bárbara Elisabeth Teixeira-Costa*[3*]

Introduction

Aromatic and medicinal plants are commonly known for their bioactive health properties, which are related to the presence of Essential Oils (EOs) in them. These substances are composed of a complex mixture of volatile compounds, such as terpenes, phenolic compounds, alcohol, constituting the secondary metabolites produced by different plant parts, roots, stems, leaves, bark, seeds, and others (Tongnuanchan and Benjakul 2014). EOs are volatile substances that are seen as colorless liquids at an ambient temperature and are highly soluble in organic solvents, alcohol, and vegetable oils, and weakly soluble in water (Falleh et al. 2020).

EOs have gained popularity over the years mainly because of their great biological activities, such as antioxidant and antimicrobial, increasing their application as natural additives in foods and food-related products. Besides, the application of these materials into food packaging has growing significantly due to the evolution of different scientific techniques and processing equipments (Tongnuanchan and Benjakul 2014). EOs displays a wide variety of flavors and scents, which are determined by their composition, origin and concentration, reaching up to 5% of the vegetable dry matter (Falleh et al. 2020, Tongnuanchan and Benjakul 2014).

[1] Graduate Program of Food Science, Center of Agrarian Sciences, Federal University of Santa Catarina–UFSC, Campus Universitário Reitor João David Ferreira Lima, 88040-900 Trindade–Florianópolis–SC, Brazil.

[2] Amend to Department of Food Technololgy, Federal University of Viçosa - UFV, Campus Universitário, Viçosa 36570-900, MG, Brazil.

[3] Amend to Department of Nutrition and Dietetics, Faculty of Nutrition Emilia de Jesus Ferreiro, Federal Fluminense University, Rua Mário Santos Braga, nº 30, 4º andar, Valonguinho, Centro - Niterói 24020-140, Niterói, Brazil..

* Corresponding author: betcosta@id.uff.br/ betcosta@gmail.com

Other factors related to cultivation practices, such as weather, use of fertilizers, rain precipitation and altitude, can influence the growth rate and the development of EO-producing plants, as well as their yield (Bhattacharya 2016). These differences have a considerable influence on the price charged by global markets.

The EOs most used in food application are derived from herbal and spice plants, such as cinnamon, clove, oregano, basil, rosemary, and others, which have been used since ancient times by civilizations, especially due to their flavor/scent and medicinal properties. Recently, consumers have been demanding novel and more natural food ingredients. In this context, the use of EOs in diverse food products has grown. The EOs to food has been considered as Generally Recognized as Safe – GRAS, although regulation agencies could require additional safety and toxicological tests prior its utilization (Mariod 2016, Mani-López et al. 2017).

As the EOs are a complex mixture of different compounds, understanding their composition is an important step to the development of novel foods with acceptable sensorial properties, as well as to their action as preserving agents. EOs can act as antioxidant and antimicrobial substances by different mechanisms, which still need more investigation, although it is known that their biological properties are linked to their content of phenolic compounds. For example, the antioxidant activity of clove, oregano, thyme and citronella EOs have been linked to their major content of active compounds eugenol, carvacrol, *m*-thymol and β-citronellol or β-citronellal, respectively (Tongnuanchan and Benjakul 2014). Despite all the benefic properties of EOs, some side effects, a strong scent, may occur when using in food products, which can negatively affect the acceptance of food by consumers (Mariod 2016). To overcome such disadvantages, some processing steps can be performed, e.g., considering indirect applications, such as in edible films and by encapsulation processes.

EOs Commonly Used in Food Matrices

Herbal and medicinal plants have a great influence on world history, because over the centuries, different civilizations used them as condiments, wound healers, perfumes, cosmetics, medicinal treatments and even as currency trading (Nascimento et al. 2020). Spices and herbs are plant-derived products, characterized by the presence of aromatic and pungent EOs that give flavor to food. Indigenous communities have used herbal plants and spices, such as black pepper and cinnamon in their foods, the Ancient Egyptians and Persian populations used them as fragrances and incense in religious ceremonies, while Greeks used pepper, cinnamon and ginger as medicines, especially for sleeping disorders, and as an additive in olive oil (Kowalski et al. 2018, Nascimento et al. 2020, Nabavi et al. 2015).

Cinnamon essential oil can be extracted from leaves, barks, and other parts of *Cinnamomum* trees. The name, cinnamon, derived from the Greek, means sweet wood (Ribeiro-Santos et al. 2017). This plant is originally from India and Sri Lanka, although nowadays, the major producers are Indonesia, followed by China, Sri Lanka, Vietnam, and Madagascar. It has been used intensively as a spice, flavoring, and preservative additive to food since ancient times. Up to 250 species of cinnamon has been identified, though only four of them have been used to obtain the spice, *C. zeylanicum* Blume or *C. verum* J. Presl (true cinnamon) from Sri Lanka, *C. aromaticum* Nees or *C. cassia* (L) J. Presl from China, *C. burmannii* (Nees & T. Nees) Blume from the Islands of Sumatra and Java, and *C. loureiroi* Ness from Vietnam (Ribeiro-Santos et al. 2017). The major components of cinnamon essential oil are cinnamaldehyde (up to 78.5% from bark) and eugenol (~76.7% from leaves), although its chemical composition can vary according to environmental and processing factors, such as the plant part used to extract the EO, type and conditions of extraction, and others (Zhou et al. 2021, Ribeiro-Santos et al. 2017). Other small components are linalool, α-pinene, limonene, β-caryophyllene, β-phellandrene, and others (Ribeiro-Santos et al. 2017). The use of cinnamon EO in food products goes beyond flavor, it is also linked to their great antimicrobial properties, e.g., *Listeria monocytogenes*, *Salmonella typhimurium, Escherichia coli*, *Bacillus cereus* and others, that are responsible for many foodborne diseases and food spoilage (Nabavi et al. 2015).

Another aromatic plant commonly used in food products and culinary practices is *Mentha* or 'mint' (*Mentha ssp.*). This plant from the Lamiaceae family is widely distributed across the world, being found in Asia, Europe, Australia, Africa, and North America (Eftekhari et al. 2021). The systematics for the *Mentha* genus is extremely complicated and far from being clarified. However, currently, four species, *M. piperita, M. canadensis, M. spicata* and *M. gracilis*, are commercially cultivated for EO production (Eftekhari et al. 2021). It is mainly used as a flavoring agent in food products, such as ice cream, liqueurs, chewing gums and candies. Besides the purposes of food, mint EO have been also used to treat other health diseases, especially due to their anti-inflammatory, anti-emetic, antispasmodic, analgesic, anticancer, anti-obesity, anti-diabetic, anti-bloating, and immunomodulatory properties (Gholamipourfard et al. 2021). *Mentha spp.* essential oil has significant importance in the global market of EOs, reaching values of US$ 400 million per annum (Eftekhari et al. 2021). In *M. piperita* leaves, the yield of essential oil can reach up to 4%, and display more than 300 compounds, which are mainly composed of monoterpenes (~52%) and sesquiterpenes (~9%) (Eftekhari et al. 2021). The major monoterpene is menthol, close to 60%, whereas the main sesquiterpene is β-caryophyllene, up to 1.8% (Eftekhari et al. 2021). Other compounds from peppermint essential oil are epoxyocimene, linalool, menthone, eucalyptol, and neo-menthol (Gholamipourfard et al. 2021). Phenolic compounds, i.e., caffeic acid and their derivatives, chlorogenic, and rosmarinic acid, are the main compounds found in *Mentha ssp.*, being used in food as natural additives due to its biological properties, especially antimicrobial and antioxidant (Eftekhari et al. 2021, Mohamed and Mansour 2012).

Basil is among the most popular herbs used for food flavoring and culinary practices. This plant belongs to the Lamiaceae family, being part of the genus *Ocimum*, which comprises more than 30 species (Suppakul et al. 2003). It is native from Asiatic countries, i.e., India, Pakistan, Iran, Thailand, and others, although it can grow wildly in tropical and sub-tropical regions of the world (Makri and Kintzios 2008). This genus is known for its wide variability of species, which result in different morphology and chemotype, leading to a diverse chemical composition of their essential oils (Silva et al. 2022, Suppakul et al. 2003). This herb and its essential oil have been used a great deal in confectionary, bakery products, condiments and sauces, such as ketchups, salad dressings, chili sauces, pickles, olive oil, pesto, vinegars, sausages and meats, as well as in nonalcoholic beverages, ice cream, and many others (Suppakul et al. 2003, Makri and Kintzios 2008). The main components of basil essential oil are terpenes and phenylpropanoids, followed by alcohols and aldehydes (Silva et al. 2022). Another recent study of Amor et al. (2021) identified the major components of basil EO as being linalool (41.3%), followed by 1,8-cineole (9.6%), (z)-isoeugenol (5.9%), 1-epi-cubenol (4.8%) and α-trans-bergamotene (4.6%), and others.

The aromatic plant rosemary (*Rosmarinus officinalis* L.) is a perennial shrub native from the Mediterranean region, belonging to the Lamiaceae family, and nowadays is distributed in all parts of the world (Silvestre et al. 2019). Rosemary has been commonly applied in culinary practices, as a food preservative, as well as in biomedical fields, such as cosmetics and in medicinal treatments for thousands of years (Ribeiro-Santos et al. 2015). The rosemary EO is colorless to a yellowish liquid that presents a camphorated scent, and is mainly composed by monoterpenes (up to 98%) and derivatives of sesquiterpenes (2–5%) (Silvestre et al. 2019). This composition can differ depending on many factors, such as climate, cultivation conditions, genetic factors and extraction forms (Silvestre et al. 2019). The major compounds found in the rosemary EO are 1,8-cineole (46.4%), camphor (11.4%) and α-pinene (11.0%), although different compositions are reported in the literature (Nieto et al. 2018). Despite antimicrobial and antioxidant properties, the rosemary EO can also display other health associated bioactivities, e.g., anti-inflammatory, antidiabetic and hepatoprotective, which can be related to its content of phenolic compounds, such as caffeic acid, rosmarinic acid and carnosic acid (Silvestre et al. 2019, Nieto et al. 2018, Ribeiro-Santos et al. 2015).

Oregano is largely cultivated in different parts of the world, and have six different botanical families with similar odor and taste characteristics, used as condiments worldwide (Leyva-López et al. 2017). Up to 61 species and 17 genera belongs to these different botanical families known as oregano. *Hedeoma patens, Lippia graveolens, Lippia palmeri, Lippia alba, Origanum dictamnus, Origanum hirtum, Origanum onites,* and *Origanum vulgare*, are species that produces EOs (Leyva-López et al. 2017). Oregano EO is mainly composed of carvacrol, thymol, cymene, and terpinene and this has been studied for application as food preservatives for their antioxidant and antimicrobial capabilities (Bedoya-Serna et al. 2018). Thyme (*Thymus vulgaris*) is another aromatic plant used for food application, e.g., meats, salads and soups, being native from southern of Europe and the Mediterranean region (Nascimento et al. 2020). Thyme essential oil is used as a natural additive to improve the flavor and aroma of food products, as well as a preservative. Its EO has more than 30 different compounds, among them being thymol (up to 10%) and carvacrol (55–85%) the main components, whereas cymene, γ-terpinene, (*E*)-caryophyllene, and linalool are also present (El-Sayed and El-Sayed 2021a). Some essential oils commonly used to food and their biological activity are listed in Table 4.1.

Table 4.1 Some essential oils commonly used to food and their biological activity

Food product	Essential oils	Plant name	Activity reported	References
Salmon and Cod fishes	Oregano oil	*Origanum vulgare*	Antimicrobial and sensorial properties	Pedrós-Garrido et al. 2020
Sea bass	Thyme oil	*Thymus vulgaris*	Antimicrobial, sensorial, and chemical properties	Kostaki et al. 2009
Minimally processed vegetables	Oregano oil, thyme oil	*Origanum vulgare, Thymus vulgaris*	Antimicrobial and sensorial properties, and quality preservation	Gutierrez et al. 2009
	Oregano oil, rosemary oil	*Origanum vulgare, Rosmarinus officinalis*	Antimicrobial properties	Azeredo et al. 2011
Fresh vegetables	Myrtle oil	*Myrtus communis*	Antimicrobial properties	Gündüz et al. 2009
Fresh-cut lettuce	Clove essential oil	*Syzygium aromaticum*	Quality preservation and browning control	Chen et al. 2017
Beef patties	Marjoram oil, rosemary oil	*Origanum majorana, Rosmarinus officinalis*	Lipid stability and sensorial properties	Mohamed and Mansour 2012
Dried meat	Oregano oil	*Origanum vulgare*	Antimicrobial and sensorial attributes	Hernández et al. 2017
White soft cheese	Cumin oil	*Cuminum cyminum*	Preservative properties	El-Sayed and El-Sayed 2021b
Cheese prepared with cream cheese base	Oregano oil, rosemary oil	*Origanum vulgare, Rosmarinus officinalis*	Preservation of sensorial attributes	Olmedo et al. 2012
Par-baked wheat and sourdough bread	Thyme oil	*Thymus vulgaris*	Antimicrobial properties	Debonne et al. 2018
Cake	Chamomile oil	*Matricaria chamomilla*	Antioxidant, and antimicrobial properties	Khaki et al. 2012
Bread	Lemongrass oil	*Cymbopogon citratus*	Antimicrobial properties	Mani-López et al. 2018
Biscuits	Cinnamon oil	*Cinnamomum zeylanicum*	Improvement of quality and stability, sensorial properties	Fadel et al. 2019
Orange juice	Orange peel oil	*Citrus sinensis*	Antimicrobial properties	Parish et al. 2003
Oil-roasted and salted peanuts	Aguaribay oil, cedron oil	*Schinus molle, Aloysia triphyla*	Antioxidant properties	Olmedo et al. 2012
Olive oil	Black pepper, fennel, orange, thyme and rosemary oils	*Piper nigrum, Foeniculum vulgare, Citrus sinensis, Thymus vulgaris,* and *Rosmarinus officinalis*	Flavoring properties	Benkhoud et al. 2021

Use of Essential Oils to Improve Functional and Preserving Properties

To Improve Sensory Attributes

Consumer acceptance is one of the most important factors to consider when developing novel products by the food industries, since it results in sales and their maintenance in a competitive market. The consumers acceptance of food products is intimately related to the study and improvement of its sensorial properties (Mani-López et al. 2017). Sensorial analysis is known as a scientific method to identify, measure, evoke, analyze and interpret products attributes by the five human senses, taste, sight, hearing, touch and smell (Mani-López et al. 2017). In this context, essential oils have been widely incorporated as additives in food products to enhance its sensorial aspects, especially adding aroma and flavor to it (Mariod 2016). When incorporating additives into food products, they should be GRAS. EOs and their constituents have been largely used as flavor/aroma ingredients in food products, since they have been considered as GRAS, although additional safety and toxicological data could be required by regulatory agencies (Mariod 2016, Mani-López et al. 2017). According to Falleh et al. (2020) since the April 1, 2018 the following EOs, cinnamon (*Cinnamomum verum*), clove (*Syzygium aromaticum*), thyme (*Thymus capitatus* – LD_{50}= >2 g kg^{-1}), rosemary (*Rosmarinus officinalis* – LD_{50}= 5.94 g kg^{-1}), mint (*Mentha cardiaca* – LD_{50}= 7.13 g kg^{-1}) and other 156 are recognized by the Code of Federal Regulations Title 21 of the United States Food and Drug Administration (FDA) as safe under the regulatory limitations. Other growing concerns related to the use of natural additives in food products, such as the EOs, is related to their contamination with organic pesticides, e.g., Ethyl chlorpyriphos and Methidathion, which were found in mandarin and lemon EOs (Falleh et al. 2020).

Generally, EOs developed and gained acceptance with countless functions due to their stimulation of human senses. When studying 12 varieties of *Ocimum basilicum*, Bernhardt et al. (2015) found that lemon scent is in close relation to the content of neral, geranial, isocitral, octanal and α-terpinene, β-cubebene in its EO. Moreover, in their work, the sweet aroma was related to the presence of furanoid, nerol, neryl-acetate and caryohyllene oxide, while the spice scent was correlated to eugenol, sabinene, β-caryophyllene, α-guiene and cis-α-bisabolene (Bernhardt et al. 2015). Thus, these authors concluded that minor components in *Ocimum basilicum* EOs could have a decisive role in consumer tests.

Fifty-five volatile compounds were identified by GC and GC–MS in the bergamot (*Citrus bergamia* Risso) essential oil produced in Reggio Calabria in Italy in the work of Sawamura et al. (2006). These authors found that the bergamot-like scent was linked to (Z)-limonene oxide, decanal, linalyl acetate and geraniol. Additionally, they estimated an aroma model of bergamot odor with the similarity of 7.1 by the nine-point-score sensory test with the mixture of eight volatile components of bergamot EO. In another study, Tsitlakidou et al. (2022) investigated the aroma profile of *Origanum vulgare* L. essential oils collected from four districts of Greece, by GC/MS and quantitative descriptive analysis using a panel of sensory specialists. These authors found that carvacrol and *p*-cymene were sensory associated with the typical scent of oregano, whereas γ-terpinene and pinene were linked to the minty and terpenic odor of oregano EOs.

The addition of spices and herbs, as well as their EOs in extra-virgin olive oil is a widespread practice in culinary practices since ancient times to produce flavored oils. Despite affecting the olive oil sensorial attributes, the addition of spices EOs also has a significant impact on their shelf-life, nutritional composition and biological activities, e.g., antioxidant (Gambacorta et al. 2007). In a recent study, Benkhoud et al. (2021) investigated the volatile profile, sensory attributes, oxidative stability, fatty acid content, and their nutritional quality of extra virgin olive oils incorporated with six different essential oils from rosemary, thyme, fennel, black pepper, Brazilian pepper tree (*Schinus terebinthifolius* L.) and orange. These authors noted that olive oils dominated by the following volatile compounds, 1,8-cineole, β-caryophyllene, trans-anethole, α-phellandrene, and limonene for thyme, rosemary, black pepper, fennel, Brazilian pepper tree, and orange, respectively. Moreover, the sensory panel analysis indicated that the olive oil flavored with fennel EO was the most preferred by consumers. Despite this, their results also indicated that the incorporation of EOs improved the oxidative stability during long-term storage and nutritional quality of the olive oil (Benkhoud et al. 2021).

To Prevent Oxidation

Some lipidic foods are susceptible to oxidation, especially those with low water activity ($aw < 0.2$) and containing high content of unsaturated fatty acids (such as oleic, linoleic, linolenic fatty acids). Due to the lipid's oxidation, the content of peroxides rises leading to the formation of numerous volatile unpleasant compounds, such as ketones and aldehydes, from rancidity indicating that the food is in a degrading process. The formation of such rancidity compounds affects not only the nutritional value of foods, but also its sensorial quality. The rancidity can occur during food processing and storage, due to the contact of unsaturated fatty acids with oxygen, mediated by catalyzers, e.g., light, metal ions, metalloproteins, enzymes and others. Antioxidant agents, including those substances present in EOs, such as phenolics, terpenes, can be used to avoid lipid oxidation and prolong the shelf life of a food (Sacchetti et al. 2005).

EOs can act as antioxidant agents through different mechanisms, e.g., preventing chain initiation, scavenging free radicals, reducing agents, terminating peroxides, preventing the continuous hydrogen abstraction, quenching the formation of singlet oxygen and binding to transition metal-ion catalysts (Tongnuanchan and Benjakul 2014). These substantial functions can be mediated by phenolic compounds present in the EOs, since they are organic compounds containing hydroxyl groups (–OH) attached directly to a carbon atom of the aromatic ring, which can be donated to free radicals, in that way preventing other compounds, such as unsaturated fatty acids, to be oxidized. This potent antioxidant activity is generally related to the major content of active compounds in EOs, e.g., eugenol in clove, carvacrol in oregano, *m*-thymol in thyme and β-citronellol or β-citronellal in citronella (Tongnuanchan and Benjakul 2014).

Other components, e.g., such alcohols, ethers, ketones, aldehydes, and monoterpenes: linalool, 1,8-Cineole, geranial/neral, citronellal, iso-menthone, menthone, can also impart significant antioxidant properties to EOs (Falleh et al. 2020). The antioxidant capacity of EOs depends on its source, harvesting period, composition, content of bioactive substances and extraction processes. Knowing the chemical composition of an EO allows an approximate prediction of its antioxidant potential. For instance, good antioxidant activity can be expected from EOs with a large amount of phenolics and little content of unsaturated terpenes (Amorati et al. 2013). Additionally, when having significant amounts of components such as cyclohexadiene, its antioxidant capacity can be even greater (Amorati et al. 2013).

EOs of oregano, cinnamon, rosemary, laurel, thyme, clove and others have been described as powerful natural antioxidants (Olmedo et al. 2012). The EOs of thyme, clove, and basil displayed relevant antioxidant activities to prevent lipidic peroxidation, which are comparable to α-tocopherol, mainly due to their content of *p*-cymene and thymol; eugenol and β-caryophyllene; and linalool, isoanethole and eugenol, respectively (Miguel 2010). The EOs of *Citrus maxima* Burm. and *Citrus sinensis* (L.) Osbeck have also shown substantial antioxidant activities, though their differences, such as limonene was the major compound in *Citrus sinensis,* while limonene, *E*-citral, *Z*-citral and 3,3-dimethyl-1-hexene were present in *Citrus maxima* (Miguel 2010).

Olmedo et al. (2012) investigated the antioxidant effect of aguaribay and cedron EOs on oil-roasted and salted peanuts. These authors noted that the addition of the aguaribay and cedron EOs had a positive effect on protecting fried-salted peanuts against lipid oxidation, which was evaluated by a reduced peroxide index, from 94.86 to 82.74 and 76.29 mEq O_2 kg^{-1}, after 112 d of storage. In another work, oregano and rosemary EOs were incorporated as preserving agents in cream cheese, and their oxidative and fermentative stabilities were investigated over 35 d of storage (Olmedo et al. 2013). In their work, Olmedo et al. (2013) noted that cream cheese incorporated with oregano and rosemary EOs showed higher stability during storage, by a lower peroxide value, 11.70 and 12.32 mEq O_2 kg^{-1}, on d 35 than the control, ~17 mEq O_2 kg^{-1}. The oxidative and microbial stability of cooked sausages incorporated with 10 and 20 ppm of nutmeg (*Myristica fragrans*) essential oil were investigated by (Šojić et al. 2015). In this work, sausages with nutmeg EO exhibited greater oxidative and microbial stability, which were visualized by lower values of thiobarbituric acid-reactive substance ranging from 1.21 mg MDA kg^{-1} and 0.95 mg MDA kg^{-1} than the control 1.53 mg MDA kg^{-1}, and lower total number of aerobic mesophilic bacteria (Šojić et al. 2015). Another study from Tomaino et al. (2005) investigated antioxidant properties and chemical composition of basil, cinnamon, clove, nutmeg, oregano and thyme EOs when heated up to 180°C. These authors noted that under room temperature all the EOs had good radical-scavenger properties by the DPPH assay, reaching an effectiveness order: clove >> cinnamon > nutmeg > basil ≥ oregano >> thyme. This behavior changed when the EOs were heated up to 180°C for 10 min, displaying the following effectiveness order: clove > thyme ≥ cinnamon > basil >> oregano > nutmeg (Tomaino et al. 2005). These results highlight the potent antioxidant capacity of EOs as natural preserving agents to be applied in different food matrices, combining their use with specific processes and methods that promote greater properties of them.

EOs can be found to be more expensive than synthetic additives when considered for food products. They can interact with some ingredients present in food formulations, giving a strong scent that can negatively affect the acceptance of food by consumers. Thus, its application should be limited to an acceptable amount and in a way that fulfills its purpose, e.g., the maximum recommended amount of berry oil in food products is about 0.025% (Mariod 2016). To overcome the side effects on using EOs into foods, some techniques can be used, such as by developing synergistic combinations with different EOs to reduce the concentration of each, applying distillation and purification processes, considering indirect applications, such as in edible films and by encapsulation processes, combining processing steps, such as low pH or refrigeration temperatures, and others.

EOs Used as Antimicrobial Agents

Food preservation combined with quality and safety maintenance, aims to increase the shelf life of products that are central points in the food production chain. At the same time, the interest for natural foods or without the presence of chemical additives by consumers are increasingly greatly. In this context, EOs can act as antimicrobial agents preventing food spoilage by extending or maintaining its shelf-life. The antimicrobial properties of different EOs have been well reported in literature by (Suppakul et al. 2003, Burt 2004, Clemente et al. 2016, Amor et al. 2021). Although EOs display greatest antimicrobial activity *in vitro*, it has been suggested that a higher content of them would be needed to reach similar properties in food systems (Burt 2004). The antimicrobial effect of EOs can be influenced by intrinsic and extrinsic factors of the food products, such as pH, salt content, storage temperature, contact with oxygen within packaging, and others. It has been reported that bacteria are more susceptible to the antimicrobial effect of EOs at lower pH of the food, storing temperature and oxygen content surrounded by packaging, and that higher fat or protein content could protect the microorganism from the EOs (Burt 2004).

The antimicrobial activities of EOs are linked a great deal to their content of phenolic compounds, which can vary depending of its origin, cultivation system and processing steps (Adelakun et al. 2016). Generally, Gram-negative bacteria are less sensitive than Gram-positive bacteria because the first group of microorganisms displays hydrophilic lipopolysaccharides in their outer membrane, allowing them superior resistance to hydrophobic compounds of essential oils (Hyldgaard et al. 2012). Low or insufficient results regarding the *in vitro* effect of terpenes, e.g., *p*-cymene, as antimicrobial agents reviewed by Hyldgaard et al. (2012) indicate that they do not have efficient antimicrobials activity when applied as single compounds. Though, terpenes that undergo biochemical modifications forming terpenoids, such as thymol, carvacrol, linalool, linalyl acetate, citronellal, piperitone, menthol, and geraniol, may have their antimicrobial properties augmented, mainly because their hydroxyl groups and the presence of delocalized electrons are important for this activity (Hyldgaard et al. 2012). Thus, terpenoids are known as a large group of antimicrobial substances with a broad activity against large spectrum of microorganisms (Hyldgaard et al. 2012).

It is known that the mode of action of these compounds occurs by the bond of amine and hydroxylamine groups of the bacterial membrane proteins, generating a change in permeability and causing its death (Adelakun et al. 2016). Although, as the EOs are a complex mixture of different substances, all the action mechanisms are not completely elucidated and it cannot be attributed to a single path.

In a study by Amor et al. (2021), the antimicrobial activity of basil essential oil against different microorganisms were investigated, aiming to produce an active chitosan-based food packaging. These authors noted that the basil EO were more active against the microorganisms of *S. salivarius, B. thermosphacta, E. coli* and *E. faecalis* than gentamicin and tetracycline. Moreover, for basil EO a minimal inhibitory concentration ranging from 1.25 to 2.50 μL/mL against *C. maltaromaticum, S. salivarius, S. saprophyticus* and *E. coli*. Thus, this EO showed effective antimicrobial activity against all microorganisms tested, both Gram-positive and Gram-negative bacteria (Amor et al. 2021).

Using the agar disk diffusion method, Kunicka-Styczyńska (2011) evaluated the effectiveness of the antimicrobial action of essential oils against nine species of yeast associated with food spoilage, namely: *Candida rugosa, Debaryomyces hansenii, Kluyveromyces marxianus, Rhodotorula glutinis, Rhodotorula minuta, Saccharomyces cerevisiae, Trichosporon cutaneum, Yarrowia lipolytica* e *Zyaccharomyces rouxigosi.* The results from this study demonstrated that all tested yeasts showed greater sensitivity to thyme oil (*Thymus vulgaris* L.) (inhibition zones of 25.5–58.0 mm). Moreover, the EOs of thyme, marjoram (*Origanum majorana* L.), peppermint (*Mentha piperita* L.) and basil (*Ocimum basilicum* L.) were the most effective (MIC and MFC values ranging from 0.5 to 24.0 μL mL^{-1}), while onion (*Allium cepa* L.) and garlic (*Allium sativum* L.) EOs were the least active and did not affect most of the strains tested.

EOs in Polymeric Packaging Films

Films and coatings act as barriers to moisture, gases and contact with the environment, preserving food products from degrading and extending its shelf-life. Edible films and coatings are mainly made by three main components, the polymer matrix, a solvent, and plasticizers (Avila-Sosa et al. 2016). The use of natural materials for the development of polymeric films, such as polysaccharides (cellulose, chitosan, alginate, carrageenan, starch and others), proteins (pea, wheat, soy, milk, gelatin, zein, etc.) and lipids (oils, waxes and resins), have been widely investigated (Teixeira-Costa and Andrade 2021). The development of bio-based packaging has been boosted by important concerns of consumers related to environment protection, reduction on the use of petroleum-based plastics which eventually end-up polluting land and sea sites, reduction of the consumption of ultra-processed foodstuffs and replacing it for more natural foods. All these concerns are related to a lifestyle changing to 'green' consumerism, which is an experiencing trend of the society, especially the Western one (Gómez-Estaca et al. 2010).

Besides this polymeric matrix, films and coatings can be developed with the incorporation of different substances, such as plasticizers, reinforcement materials, aroma and color additives, as well as preserving agents, such as synthetic antioxidant compounds like BHA (butylhydroxyanisole) or BHT (butylhydroxytoluene), and essential oils (Simona et al. 2021). Nowadays, there is growing interest in the use of natural ingredients, i.e. EOs, as an alternative to those of synthetic origin, i.e., BHA and BHT, which have been associated with certain health risks (Salgado et al. 2015).

The incorporation of active agents can be conducted inside or onto the surface of the biopolymeric films (Salgado et al. 2015). The design of novel active packaging is an increasingly dynamic technology for food preservation, mainly because this material can positively interact with the food product and can gradually shift to it. More recently, some intelligent packaging has been developed aiming to sense, detect, and record internal or external changes in food and in the environment as well (Fang et al. 2017). Intelligent packaging is largely related to active packaging, since the first sense and measures a food attribute, while the second has active functions within the product (Sohail et al. 2018). Thus, the incorporation of diverse composite materials can influence the physico-chemical characteristics as well as the functionality of the films formed.

This area of expertise has been growing rapidly, especially over the last 5 yr. When searching for publications using the keywords, edible films, and essential oils, in scientific platforms, i.e., Web of Science, from 2000 to 2022, a major increase was noted in published papers after 2015. The result of this is showed in Figure 4.1.

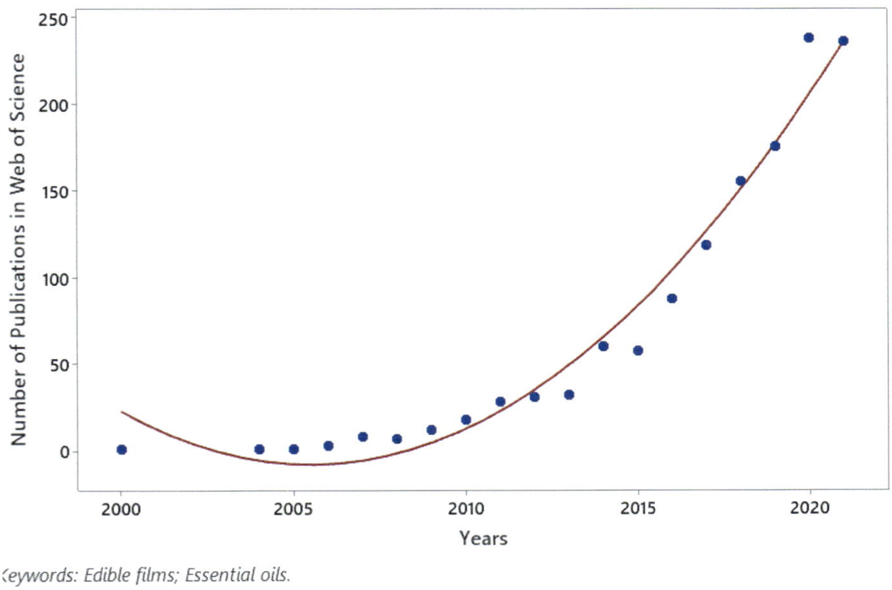

Figure 4.1 Scatter plot related to the number of published papers in Web of Science® from 2000 to 2021 with the keywords, edible films, and essential oils.

A total of 1,307 paper were published in Web of Science® (WoS) over the last 21 yr related to edible films and essential oils. It is possible to note a higher increase in publications occurring after 2010. It is also possible to suggest that in 2022 a similar or greater number of publications related to this area than in the year of 2021 ($\geq$ 236 papers) is also expected. A quadratic regression was plotted according to Equation 1 and a well-fitted determination coefficient ($R^2 = 96.9$) is presented.

$$\text{Publications in WoS} = 4084206 - 4073\ \text{Years} + 1.015\ \text{Years}^2 \tag{1}$$

When looking only at the years from 2020 and 2021, it should be noted that the number of publications was similar in both years, around 237 papers, which could be related to the COVID-19 Pandemic, mainly because of movement restrictions that limited the access to investigation laboratories and affected many scientific studies all over the world, while many researchers have directed their efforts to study and develop vaccines/medicines against COVID-19 (Palayew et al. 2020). Moreover, this data gives an important overview of the research area related to the incorporation of essential oils in edible films to diverse purposes.

Advantage and Application

Biopolymeric films or coatings incorporated with EOs has been used in food industries with a wide different purpose, such as to confer antimicrobial and antioxidant activities, to improve sensorial, functional, rheological, and barrier properties. EOs can be encapsulated before the addition or directly incorporated into the films and coatings (Benbettaïeb et al. 2017). These bioactive EOs-enriched films can be used for the preservation of fruits, vegetables, meat, fish, dairy products, and others (Salgado et al. 2015). Despite these functions, edible coatings and films enriched with EOs can display other properties, such as prevent fat and lipids oxidation, retain aroma and other volatile compounds, delay or prevent moisture migration, as well as control gases exchange, i.e., slowing respiration of many fruits and vegetables (Avila-Sosa et al. 2016).

One of the major advantages of using EOs into polymeric films or coatings is related to the controlled release of the bioactive agent through a slow diffusion rate, which keeps the active substances concentrated on the food product surface for longer periods and effectively reduces the microbial growth (Sánchez-González et al. 2011). The migration of EOs to the product surface depends of different factors, generally linked to the physico-chemical interactions of it with the polymeric matrix, such as electrostatic interaction, ionic osmoses, diffusion, water activity, film composition, pH, and other environmental conditions in which the films could be subjected to temperature and storage time (Avila-Sosa et al. 2016). The incorporation of essential oils has the advantage of using small amounts of the bioactive additive that will have a positive antimicrobial effect, as they have a low diffusion rate through the film, slowly release onto the food surface (Gómez-Estaca et al. 2010, Avila-Sosa et al. 2016).

Edible coatings and edible films are similar in their function as packaging materials but are different in the application method. The first are solid laminates, which can be applied to the foodstuff surface as a thin membrane, whereas the second are applied in liquid methods, as filmogenic solutions, in which the food is increased in it (Mohamed et al. 2020, Teixeira-Costa and Andrade 2021). The composite matrix

Table 4.2 Some films and coatings incorporated with EOs used to antimicrobial properties

Film matrix	Essential Oil	Microorganisms	References
Carrageenan	Orange	*Staphylococcus aureus* subsp. *aureus, Escherichia coli,* and *Candida albicans*	Simona et al. 2021
Artemisia sphaerocephala Krasch. gum	Star anise	Total viable aerobic bacteria and *E. coli*	Liu et al. 2020
Whey protein	Lemon and bergamot	*E. coli, S. aureus,* and *Aspergillus niger*	Çakmak et al. 2020
	Cinnamon (*C. cassia, C. zeylanicum*), rosemary and basil	*E. coli, S. aureus* and *Penicillium spp.*	Ribeiro-Santos et al. 2017
Sodium alginate	*M. pulegium* L., *A. herba alba* Asso, *O. basilicum* L. and *R. officinalis* L.	*S. aureus, E. coli, Salmonella enterica, Enterococcus faecium, Klebsiella pneumoniae* and *Enterococcus faecalis*	Mahcene et al. 2020
	Lemongrass	*E. coli* and *Listeria monocytogenes*	Bustos et al. 2016
	Thyme, lemongrass, or sage	*E. coli*	Acevedo-Fani et al. 2015
Soy protein isolate-gum acacia composite	Oregano, lemon, fruit of *Amomum tsaoko* Crevost et Lemaire or grapefruit	*E. coli* and *Bacillus subtilis*	Xue et al. 2019
Soluble soybean polysaccharide	Cinnamon	*S. aureus* and *Streptococcus pyogenes*	Ghani et al. 2018
Basil seed gum	Oregano	*Salmonella Typhimurium, Bacillus cereus, E. coli, S. aureus* and *Pseudomonas aeruginosa*	Hashemi et al. 2017
Chitosan	*Eucalyptus globulus*	*E. coli, S. aureus, P. aeruginosa, Candida albicans and Candida parapsilosis*	Hafsa et al. 2016
	Basil	*E. coli* and *S. saprophyticus*	Amor et al. 2021
Citrus pectin	Clove bud	*E. coli, S. aureus* and *L. monocytogenes*	Nisar et al. 2018
Cassava starch composite	Cinnamon	*Penicillium commune* and *Eurotium amstelodami*	Souza et al. 2013
Chitosan and hydroxypropyl methylcellulose	Bergamot, lemon, or tea tree	*E. coli, Listeria monocytogenes* and *S. aureus*	Sánchez-González et al. 2011
Gelatine-chitosan	Clove, fennel, cypress, lavender, thyme, herb-of-the-cross, pine and rosemary	*E. coli, Pseudomonas fluorescens, Lactobacillus acidophilus* and *Listeria innocua*	Gómez-Estaca et al. 2010

of coatings and films can be obtained as emulsions or bilayers, although this last is rarely used in food industries, as they need two casting and two drying stages, even with providing good barrier properties (Mohamed et al. 2020).

Antimicrobial compounds naturally found in EOs are widely distributed in the environment and have been studied as preservation systems in diverse food films and coatings. The physical and antimicrobial properties of sodium alginate/carboxymethyl cellulose composite films incorporated with cinnamon essential oil were investigated by Han et al. 2018. These authors found that the incorporation of cinnamon EO affected the physical characteristics of the produced films, such as by increasing the thickness, water-vapor permeability, oxygen permeability, and elongation at break, and decreasing the moisture content and tensile strength. The sodium alginate/carboxymethyl cellulose cinnamon essential oil composite films exhibited excellent antimicrobial activity for *Escherichia coli* and *Staphylococcus aureus* (Han et al. 2018). In another study, the incorporation of various concentrations of cinnamon EO in fish skin gelatin films were investigated by Wu et al. (2017). In this work, the gelatin-cinnamon-EO films exhibited greater antifungal activity, which was more effective than the antibacterial, although the films had good inhibitory effects against all tested microorganisms, i.e., *Escherichia coli, Staphylococcus aureus, Aspergillus niger, Rhizopus oryzae*, and *Paecilomyces varioti*. The composition of some films and coatings incorporated with EOs due to its antimicrobial properties are listed in Table 4.2.

In spite of being used as antimicrobial and/or antioxidant agents, EOs can also affect the structure and function of bio-based films, due to their surfactant and emulsification properties (Benbettaïeb et al. 2017). Chitosan films contain from 0–4% of *Eucalyptus globulus* essential oil were studied by Hafsa et al. (2016). These authors noted that when the EO concentration increased, the moisture and solubility of the films significantly decreased, which can be beneficial properties for the application of food products. Wu et al. (2017) observed that the tensile strength, elongation at break, and water content of the gelatin-cinnamon-EO based films decreased with increasing cinnamon-EO concentration, while the water vapor permeability increased.

Conclusions

EOs are composed of a complex mixture of volatile compounds, such as terpenes and phenolic compounds, that constitute secondary metabolites produced by different plant parts, from where they can be extracted for different diverse bioactive applications, such as in food products. As natural food additives, EOs can be used to prevent lipidic oxidation, extend the shelf-life, inhibit microbiological growth, and incorporate into bio-based packaging of foods. The EOs are alternatives to synthetic chemicals additives (BHT and BHA), being less harmful to human health and more environmentally friendly. Thus, they are the greatest alternatives as natural preserving agents to diverse food application.

References

Acevedo-Fani, A., L. Salvia-Trujillo, M.A. Rojas-Graü and O. Martín-Belloso. 2015. Edible films from essential-oil-loaded nanoemulsions: physicochemical characterization and antimicrobial properties. Food Hydrocolloids. 47 (May): 168–177.

Adelakun, O.E., O.J. Oyelade and B.F. Olanipekun. 2016. Use of essential oils in food preservation. pp. 71–84. *In:* Victor R. Preedy (ed.). Essential Oils in Food Preservation, Flavor and Safety. Elsevier Inc.

Amor, G., M. Sabbah, L. Caputo, M. Idbella, V. De Feo, R. Porta, et al. 2021. Basil essential oil: Composition, antimicrobial properties, and microencapsulation to produce active chitosan films for food packaging. Foods. 10(1): 121.

Amorati, R., M.C. Foti and L. Valgimigli. 2013. Antioxidant activity of essential oils. J. Agric. Food Chem. 61(46): 10835–10847.

Avila-Sosa, R., E. Palou and A. López-Malo. 2016. Essential oils added to edible films. pp. 149–154. *In:* Victor R. Preedy (ed.). Essential Oils in Food Preservation, Flavor and Safety. Elsevier Inc.

Azeredo, G.A., T.L.M. Stamford, P.C. Nunes, N.J. Gomes Neto, M.E.G. de Oliveira and E.L. de Souza. 2011. Combined application of essential oils from *Origanum vulgare* L. and *Rosmarinus officinalis* L. to inhibit bacteria and autochthonous microflora associated with minimally processed vegetables. Food Res. Int. 44(5): 1541–1548.

Bedoya-Serna, C.M., G.C. Dacanal, A.M. Fernandes and S.C. Pinho. 2018. Antifungal activity of nanoemulsions encapsulating oregano (*Origanum vulgare*) essential oil: *In vitro* study and application in Minas Padrão cheese. Braz. J. Microbiol. 49(4): 929–935.

Benbettaïeb, N., T. Karbowiak and F. Debeaufort. 2017. Bioactive edible films for food applications: Influence of the bioactive compounds on film structure and properties. Crit. Rev. Food Sci. Nutr. 8398: 1–17.

Benkhoud, H., Y. M'Rabet, M. Gara ali, M. Mezni and K. Hosni. 2021. Essential oils as flavoring and preservative agents: impact on volatile profile, sensory attributes, and the oxidative stability of flavored extra virgin olive oil. J. Food Process. Preserv. 46(5): e15379.

Bernhardt, B., L. Sipos, Z. Kókai, A. Gere, K. Szabó, J. Bernáth, et al. 2015. Comparison of different *Ocimum basilicum* L. gene bank accessions analyzed by GC–MS and sensory profile. Ind. Crops Prod. 67: 498–508.

Bhattacharya, S. 2016. Cultivation of essential oils. pp. 19–29. *In:* Victor R. Preedy (ed.). Essential Oils in Food Preservation, Flavor and Safety. Elsevier Inc.

Burt, S. 2004. Essential oils: Their antibacterial properties and potential applications in foods — a review. Int. J. Food Microbiol. 94: 223–253.

Bustos, C, R.O., F.V. Alberti R. and S.B. Matiacevich. 2016. Edible antimicrobial films based on microencapsulated lemongrass oil. J. Food Sci. Technol. 53(1): 832–839.

Çakmak, H., Y. Özselek, O.Y. Turan, E. Fıratlıgil and F. Karbancioğlu-Güler. 2020. Whey protein isolate edible films incorporated with essential oils: Antimicrobial activity and barrier properties. Polym. Degrad. Stab. 179: 109285.

Chen, X., L. Ren, M. Li, J. Qian, J. Fan and B. Du. 2017. Effects of clove essential oil and eugenol on quality and browning control of fresh-cut lettuce. Food Chem. 214: 432–439.

Clemente, I., M. Aznar, F. Silva and C. Nerín. 2016. Antimicrobial properties and mode of action of mustard and cinnamon essential oils and their combination against foodborne bacteria. Innovative Food Sci. Emerging Technol. 36: 26–33.

Debonne, E., F. Van Bockstaele, I. De Leyn, F. Devlieghere and M. Eeckhout. 2018. Validation of *in-vitro* antifungal activity of thyme essential oil on *Aspergillus niger* and *Penicillium paneum* through application in par-baked wheat and sourdough bread. LWT. 87: 368–378.

Eftekhari, A., A. Khusro, E. Ahmadian, S.M. Dizaj, A. Hasanzadeh and M. Cucchiarini. 2021. Phytochemical and nutra-pharmaceutical attributes of mentha Spp.: A comprehensive review. Arabian J. Chem. 14(5): 103106.

El-Sayed, S.M. and H.S. El-Sayed. 2021a. Antimicrobial nanoemulsion formulation based on thyme (*Thymus vulgaris*) essential oil for UF labneh preservation. J. Mater. Res. Technol. 10: 1029–1041.

El-Sayed, H.S. and S.M. El-Sayed. 2021b. A modern trend to preserve white soft cheese using nano-emulsified solutions containing cumin essential oil. Environ. Nanotechnol., Monit. Manage. 16: 100499.

Fadel, H.H.M., I.M. Hassan, M.T. Ibraheim, M.A. Abd El Mageed and R. Saad. 2019. Effect of using cinnamon oil encapsulated in maltodextrin as exogenous flavouring on flavour quality and stability of biscuits. J. Food Sci. Technol. 56(10): 4565–4574.

Falleh, H., M. Ben Jemaa, M. Saada and R. Ksouri. 2020. Essential oils: A promising eco-friendly food preservative. Food Chem. 330: 127268.

Fang, Z., Y. Zhao, R.D. Warner and S.K. Johnson. 2017. Active and intelligent packaging in meat industry. Trends Food Sci. Technol. 61(2): 60–71.

Gambacorta, G., M. Faccia, S. Pati, C. Lamacchia, A. Baiano and E. La Notte. 2007. Changes in the chemical and sensorial profile of extra virgin olive oils flavored with herbs and spices during storage. J. Food Lipids. 14(2): 202–215.

Ghani, S., H. Barzegar, M. Noshad and M. Hojjati. 2018. The preparation, characterization and *in vitro* application evaluation of soluble soybean polysaccharide films incorporated with cinnamon essential oil nanoemulsions. Int. J. Biol. Macromol. 112: 197–202.

Gholamipourfard, K., M. Salehi and E. Banchio. 2021. Mentha piperita phytochemicals in agriculture, food industry and medicine: Features and applications. S. Afr. J. Bot. 141: 183–195.

Gómez-Estaca, J., A. López de Lacey, M.E. López-Caballero, M.C. Gómez-Guillén and P. Montero. 2010. Biodegradable gelatin-chitosan films incorporated with essential oils as antimicrobial agents for fish preservation. Food Microbiol. 27(7): 889–896.

Gündüz, G.T., Ş.A. Gönül and M. Karapinar. 2009. Efficacy of myrtle oil against *Salmonella* Typhimurium on fresh produce. Int. J. Food Microbiol. 130(2): 147–150.

Gutierrez, J., P. Bourke, J. Lonchamp and C. Barry-Ryan. 2009. Impact of plant essential oils on microbiological, organoleptic and quality markers of minimally processed vegetables. Innovative Food Sci. Emerging Technol. 10(2): 195–202.

Hafsa, J., M. ali Smach, M.R. Ben Khedher, B. Charfeddine, K. Limem, H. Majdoub, et al. 2016. Physical, antioxidant and antimicrobial properties of chitosan films containing eucalyptus globulus essential oil. LWT 68: 356–364.

Han, Y., M. Yu and L. Wang. 2018. Physical and antimicrobial properties of sodium alginate/carboxymethyl cellulose films incorporated with cinnamon essential oil. Food Packag. Shelf-Life. 15: 35–42.

Hashemi, S.M.B. and A. Mousavi Khaneghah. 2017. Characterization of novel basil-seed gum active edible films and coatings containing oregano essential oil. Prog. Org. Coat. 110: 35–41.

Hernández, H., A. Fraňková, T. Sýkora, P. Klouček, L. Kouřimská, I. Kučerová and J. Banout. 2017. The effect of oregano essential oil on microbial load and sensory attributes of dried meat. J. Sci. Food Agric. 97(1): 82–87.

Hyldgaard, M., T. Mygind and R.L. Meyer. 2012. Essential oils in food preservation: Mode of action, synergies, and interactions with food matrix components. Front. Microbiol. 3: 12.

Khaki, M., M. Sahari and M. Barzegar. 2012. Evaluation of antioxidant and antimicrobial effects of Chamomile (*Matricaria chamomilla* L.) essential oil on cake shelf. J. Med. Plants. 11(43): 9–18.

Kostaki, M., V. Giatrakou, I.N. Savvaidis and M.G. Kontominas. 2009. Combined effect of MAP and thyme essential oil on the microbiological, chemical and sensory attributes of organically aquacultured sea bass (*Dicentrarchus labrax*) Fillets. Food Microbiol. 26(5): 475–482.

Kowalski, R., G. Kowalska, U. Pankiewicz, A. Mazurek, M. Sujka, M. Włodarczyk-Stasiak, et al. 2018. Effect of the method of rapeseed oil aromatisation with rosemary *Rosmarinus officinalis* L. on the content of volatile fraction. LWT. 95: 40–46.

Kunicka-Styczyńska, A. 2011. Activity of essential oils against food-spoiling yeast. A review. Flavour Fragrance J. 26(5): 326–328.

Leyva-López, N., E. Gutiérrez-Grijalva, G. Vazquez-Olivo and J. Heredia. 2017. Essential oils of oregano: Biological activity beyond their antimicrobial properties. Molecules. 22(6): 989.

Liu, Q., M. Zhang, B. Bhandari, J. Xu and C. Yang. 2020. Effects of nanoemulsion-based active coatings with composite mixture of star anise essential oil, polylysine, and nisin on the quality and shelf life of ready-to-eat yao meat products. Food Control. 107: 106771.

Mahcene, Z., A. Khelil, S. Hasni, P.K. Akman, F. Bozkurt, K. Birech, et al. 2020. Development and characterization of sodium alginate based active edible films incorporated with essential oils of some medicinal plants. Int. J. Biol. Macromol. 145: 124–132.

Makri, O. and S. Kintzios. 2008. *Ocimum* Sp. (Basil): Botany, cultivation, pharmaceutical properties, and biotechnology. J. Herbs, Spices Med. Plants. 13(3): 123–150.

Mani-López, E., A.C. Lorenzo-Leal, E. Palou and A. López-Malo. 2017. Principles of sensory evaluation in foods containing essential oil. pp. 293–325. *In:* S.M.B. Hashemi, A.M. Khaneghah and A. de S. Sant'Ana (eds). Essential Oils in Food Processing: Chemistry, Safety and Applications. John Wiley & Sons Ltd.

Mani-López, E., G.P. Valle Vargas, E. Palou and A. López Malo. 2018. *Penicillium expansum* inhibition on bread by lemongrass essential oil in vapor phase. J. Food Prot. 81(3): 467–471.

Mariod, A.A. 2016. Effect of essential oils on organoleptic (Smell, Taste, and Texture) properties of food. pp. 131–137. *In:* Victor R. Preedy (ed.). Essential Oils in Food Preservation, Flavor and Safety. Elsevier Inc.

Miguel, M.G. 2010. Antioxidant and anti-inflammatory activities of essential oils: A short review. Molecules. 15(12): 9252–9287.

Mohamed, H.M.H. and H.A. Mansour. 2012. Incorporating essential oils of marjoram and rosemary in the formulation of beef patties manufactured with mechanically deboned poultry meat to improve the lipid stability and sensory attributes. LWT. 45(1): 79–87.

Mohamed, S.A.A., M. El-Sakhawy and M.A.-M. El-Sakhawy. 2020. Polysaccharides, protein and lipid-based natural edible films in food packaging: A review. Carbohydr. Polym. 238: 116178.

Nabavi, S., A. Di Lorenzo, M. Izadi, E. Sobarzo-Sánchez, M. Daglia and S. Nabavi. 2015. Antibacterial effects of cinnamon: From farm to food, cosmetic and pharmaceutical industries. Nutrients. 7(9): 7729–7748.

Nascimento, L.D. do, A.A.B. de Moraes, K.S. da Costa, J.M. Pereira Galúcio, P.S. Taube, C.M.L. Costa, et al. 2020. Bioactive natural compounds and antioxidant activity of essential oils from spice plants: New findings and potential applications. Biomolecules. 10(7): 988.

Nieto, G., G. Ros and J. Castillo. 2018. Antioxidant and antimicrobial properties of rosemary (*Rosmarinus officinalis*, L.): A review. Medicines. 5(3): 98.

Nisar, T., Z.C. Wang, X. Yang, Y. Tian, M. Iqbal and Y. Guo. 2018. Characterization of citrus pectin films integrated with clove bud essential oil: Physical, thermal, barrier, antioxidant and antibacterial properties. Int. J. Biol. Macromol. 106: 670–680.

Olmedo, R.H., V. Nepote and N.R. Grosso. 2012. Aguaribay and cedron essential oils as natural antioxidants in oil-roasted and salted peanuts. J. Am. Oil Chem. Soc. 89(12): 2195–2205.

———. 2013. Preservation of sensory and chemical properties in flavoured cheese prepared with cream cheese base using oregano and rosemary essential oils. LWT. 53(2): 409–417.

Palayew, A., O. Norgaard, K. Safreed-Harmon, T.H. Andersen, L.N. Rasmussen and J. V. Lazarus. 2020. Pandemic publishing poses a new COVID-19 challenge. Nat. Human Behav. 4(7): 666–669.

Parish, M.E., D. Baum, R. Kryger, R. Goodrich and R. Baum. 2003. Fate of salmonellae in citrus oils and aqueous aroma. J. Food Prot. 66(9): 1704–1707.

Pedrós-Garrido, S., I. Clemente, J.B. Calanche, S. Condón-Abanto, J.A. Beltrán, J.G. Lyng, et al. 2020. Antimicrobial activity of natural compounds against *Listeria spp.* and their effects on sensory attributes in salmon (*Salmo salar*) and cod (*Gadus morhua*). Food Control. 107: 106768.

Ribeiro-Santos, R., M. Andrade, D. Madella, A.P. Martinazzo, L. de Aquino Garcia Moura, N.R. de Melo, et al. 2017. Revisiting an ancient spice with medicinal purposes: Cinnamon. Trends Food Sci. Technol. 62: 154–169.

Ribeiro-Santos, R., D. Carvalho-Costa, C. Cavaleiro, H.S. Costa, T.G. Albuquerque, M.C. Castilho, et al. 2015. A novel insight on an ancient aromatic plant: The rosemary (*Rosmarinus officinalis* L.). Trends Food Sci. Technol. 45(2): 355–368.

Ribeiro-Santos, R., A. Sanches-Silva, J.F.G. Motta, M. Andrade, I. de A. Neves, R.F. Teófilo, et al. 2017. Combined use of essential oils applied to protein base active food packaging: Study *in vitro* and in a food simulant. Eur. Polym. J. 93: 75–86.

Sacchetti, G., S. Maietti, M. Muzzoli, M. Scaglianti, S. Manfredini, M. Radice, et al. 2005. Comparative evaluation of 11 essential oils of different origin as functional antioxidants, antiradicals and antimicrobials in foods. Food Chem. 91(4): 621–632.

Salgado, P.R., C.M. Ortiz, Y.S. Musso, L. Di Giorgio and A.N. Mauri. 2015. Edible films and coatings containing bioactives. Curr. Opin. Food Sci. 5: 86–92.

Sánchez-González, L., M. Cháfer, M. Hernández, A. Chiralt and C. González-Martínez. 2011. Antimicrobial activity of polysaccharide films containing essential oils. Food Control. 22(8): 1302–1310.

Sawamura, M., Y. Onishi, J. Ikemoto, N.T.M. Tu and N.T.L. Phi. 2006. Characteristic odour components of bergamot (*Citrus bergamia* Risso) essential oil. Flavour Fragrance J. 21(4): 609–615.

Silva, W.M.F. da, D.H. Kringel, E.J.D. de Souza, E. da Rosa Zavareze and A.R.G. Dias. 2022. Basil essential oil: Methods of extraction, chemical composition, biological activities, and food applications. Food Bioprocess Technol. 15(1): 1–27.

Silvestre, W.P., F.R. Medeiros, F. Agostini, D. Toss and G.F. Pauletti. 2019. Fractionation of rosemary (*Rosmarinus officinalis* L.) essential oil using vacuum fractional distillation. J. Food Sci. Technol. 56(12): 5422–5434.

Simona, J., D. Dani, S. Petr, N. Marcela, T. Jakub and T. Bohuslava. 2021. Edible films from carrageenan/orange essential oil/trehalose—structure, optical properties, and antimicrobial activity. Polymers. 13(3): 332.

Sohail, M., D.-W. Sun and Z. Zhu. 2018. Recent developments in intelligent packaging for enhancing food quality and safety. Crit. Rev. Food Sci. Nutr. 58(15): 2650–2662.

Šojić, B., V. Tomović, S. Kocić-Tanackov, S. Škaljac, P. Ikonić, N. Džinić, et al. 2015. Effect of nutmeg (*Myristica fragrans*) essential oil on the oxidative and microbial stability of cooked sausage during refrigerated storage. Food Control. 54: 282–286.

Souza, A.C., G.E.O. Goto, J.A. Mainardi, A.C. V Coelho and C.C. Tadini. 2013. Cassava starch composite films incorporated with cinnamon essential oil: Antimicrobial activity, microstructure, mechanical and barrier properties. LWT. 54(2): 346–352.

Suppakul, P., J. Miltz, K. Sonneveld and S.W. Bigger. 2003. Antimicrobial properties of basil and its possible application in food packaging. J. Agric. Food Chem. 51(11): 3197–3207.

Teixeira-Costa, B.E. and C.T. Andrade. 2021. Natural polymers used in edible food packaging—History, function and application trends as a sustainable alternative to synthetic plastic. Polysaccharides. 3(1): 32–58.

Tomaino, A., F. Cimino, V. Zimbalatti, V. Venuti, V. Sulfaro, A. De Pasquale, et al. 2005. Influence of heating on antioxidant activity and the chemical composition of some spice essential oils. Food Chem. 89(4): 549–554.

Tongnuanchan, P. and S. Benjakul. 2014. Essential oils: Extraction, bioactivities, and their uses for food preservation. J. Food Sci. 79(7): R1231–R1249.

Tsitlakidou, P., A. Papachristoforou, N. Tasopoulos, A. Matzara, M. Hatzikamari, K. Karamanoli, et al. 2022. Sensory analysis, volatile profiles and antimicrobial properties of *Origanum vulgare* L. essential oils. Flavour Fragrance J. 37(1): 43–51.

Wu, J., X. Sun, X. Guo, S. Ge and Q. Zhang. 2017. Physicochemical properties, antimicrobial activity and oil release of fish gelatin films incorporated with cinnamon essential oil. Aquacult. Fish. 2(4): 185–192.

Xue, F., Y. Gu, Y. Wang, C. Li and B. Adhikari. 2019. Encapsulation of essential oil in emulsion based edible films prepared by soy protein isolate-gum acacia conjugates. Food Hydrocolloids. 96: 178–189.

Zhou, Y., X. Wu, J. Chen and J. He. 2021. Effects of cinnamon essential oil on the physical, mechanical, structural and thermal properties of cassava starch-based edible films. Int. J. Biol. Macromol. 184: 574–583.

5

Essential Oils in Food and Agriculture: Advances, Limitations and Future Prospects

Peter Oluyori Abimbola[1,2] *and Eunice Olanipekun Bolatito*[3*]

Introduction

Agriculture has been the means of livelihood for many in both rural and urban areas. It produces material for shelter, medicinal plants, and global food. Several challenges/disasters affect the productiveness, effectiveness, determination and yields of agricultural products. Part of these challenges include; long term storage of agricultural products, weed control, pest control, microbial food spoilage, etc. At the same time, the recent outburst of foodborne diseases caused by microbes such as; *Listeria monocytogenes*, *Salmonella Saintpaul*, *Escherichia coli* O157:H7 prioritized microbiological safety of agricultural products, especially, during food storage and processing (Zhang et al. 2017). In finding a solution to these challenges, several synthetic chemicals have been produced and tested on several agricultural produce, both in the process of making and the final products. These synthetic chemicals are either toxic to humans and the environment, or are affected by developing resistance to them by pests. In search for safer solutions to these agricultural issues, Essential Oils (EOs) came to use.

Essential oils are volatile secondary metabolites of aromatic plants which are usually developed for their fragrance, flavour, bio—and pharmacological use. Essential oils and their biologically active compounds have become the preferred alternatives to synthetic chemicals used as pesticides (herbicides, fungicides, and insecticides) in agriculture and for their antimicrobial efficacy against food borne pathogens (related toxins, moulds and bacteria) in the food sectors due to their ecofriendliness, and safety/low toxicity.

[1] Landmark University SDG 3 (Good health and well-being Research Group).
[2] Department of Physical Sciences, Landmark University, Kwara State, Nigeria.
[3] Department of Chemistry & Industrial Chemistry, Kwara state University, PMB 1530, Malete, Kwara state, Nigeria.
* Corresponding author: Aunice Bolatito Pekuntito@gmail.com

Essential Oils as Pesticides (Herbicides, Fungicides, and Insecticides)

Pesticides are a heterogenous group of chemicals with a significant benefit to the public since they antogonize food-borne- and vector-borne diseases thereby increasing food productivity, that is, reduce postharvest losses and increase crop yield (Weiss et al. 2004). Meanwhile, their expanded use may result in residues in foods which could pose health risks depending on the exposure and agent (Al-Saleh 1994). The quests for safer pesticides have lead researchers to make use of essential oils.

EOs extracted from various aromatic plants have been reported to pose no hazard to humans and the environment, thus, encouraging their use as pesticides (Table 5.1). For instance, *Ficus exasperata* Vahl. (Moraceae), also known as sandpaper are dominant in African countries such as; Senegal, Nigeria, Ethiopia, Mozambiq, etc. and Southern Asia, viz., Sri Lanka, India, and Yemen). The EOs of leaves of *Ficus exasperate* proves to have acaricidal activity against *Rhipicephalus sanguineus* (dog ticks) (Oyagbemi et al. 2019). Oladosu et al. 2009, also reported the antifungal activity of the plant EO against *Candida albicans*.

Leptospermum scoparium (Myrtaceae), otherwise called New Zealand tea tree or manuka is a flowering plant species commonly grown in South-East Australia and New Zealand. Duque et al. 2021 documented the acaricidal property of EO of *Leptospermum scoparium* prepared using ethanol, DMSO and tween 80 against larva and adult *Rhipicephalus microplus*. They reported 100% mortality of the larva at all fraction concentrations (0.625–10 mg/mL) tested and concluded that the EO of *Leptospermum scoparium* as well as the fractions derived from it have an acaricidal property. The virucidal activity of this oil against Herpes simplex virus type 1 and 2 gave a potency with IC_{50} of 0.96 and 0.58 μg/mL respectively (Reichling et al. 2005). *Origanum vulgare* (Lamiaceae), also known as oregano, is commonly grown in the Mediterranean region, predominant in Turkey (Arabaci et al. 2020).

Essential Oils Possessing Antimicrobial Activity

Antimicrobial agents are chemicals that helps the prevention and inhibition of the growth of microorganisms. Food antimicrobial agents are usually used together with other preservatives to prolong the shelf life of food. Usually, chemically synthesized preservatives are being used in the food industry (Prakash et al. 2018). The concern for the potential hazard such as mutagenic and carcinogenic risks associated with the chemical preservatives increases as consumers' awareness increases, thereby increasing the urge and demand for more natural and minimally processed food. In view of this, a lot of attention has been given to the search and development of less hazardous antimicrobial agents in replacement of the synthetic ones. This brings essential oils into the context. Many EOs have been tested against various microorganisms as summarized in Table 5.2.

Table 5.1 Name and constituents of essential oils used as acaricide, insecticides and fungicides

S. N.	EO	Constituents	Remarks	PEST	References
	ACARICIDAL PROPERTY				
1.	Lemon grass, moutan bark, citronella, and patchouli essential oils.	Essential oils + propylene glycol, C12 amino sucrose ester, PEG-40 hydrogenated castor oil and the balance water. **Release control**	0.02–0.04% of lemon grass essential oil, 0.04–0.08% of moutan bark essential oil, 0.01–0.03% of citronella essential oil, 0.01–0.03% of patchouli essential oil, 2–5% of propylene glycol, 2–5% of C12 amino sucrose ester, 1–2% of PEG-40 hydrogenated castor oil, and the balance of water.		Zhao et al. 2021
2.	*Ficus exasperata*	Citronellyl propionate, Neophytadiene, Lavandulyl acetate, Octadecanoic acid Me ester, Tridecanoic acid Yadanzioside, Pentanoic acid, 4-methyl- (Methylvaleric acid)	100% mortality of *Rhipicephalus sanguineus* larvae at 0.15 mg/mL of *n*-hexane ext. and 0.24 mg/mL of methanol ext. of FE.	*Rhipicephalus sanguineus* (dog ticks) 100% mortality of the adult *Rhipicephalus sanguineus* by 0.15 mg/mL of *n*-hexane extract and 98% mortality was observed at 0.18 mg/mL of methanol ext. LD (LD_{50}) of *Ficus exasperata* is 0.12.	Oyagbemi et al. 2019
3.	*Litsea cubeba, Origanum vulgare* and *Leptospermum scoparium*	geranial, carvacrol and cis-calamenene	Acaricidal activity of these EOs in Solutions. Prepared using ethanol, dimethylsulfoxide (DMSO) and Tween 80 on larvae and females of *R. microplus*. 2.5 to 10.0 mg/mL.	*Rhipicephalus microplus* 100% mortality of larvae in treatments with *L. scoparium* and *O. vulgare* EO (dild. both in DMSO and in ethanol).	Duque et al. 2021
4.	Turmeric volatile oil: *Rhizoma Curcumae* Longae (Curcuma longa L.)		Mortality rate within 48 hr was more than 80%, and significantly reduced the nematode activity.	Pine wood nematode (Bursaphelenchusxylophilus).	Li et al. 2021 **PATENT**
5.	*Lippia alba*	Geraniol neral	1500 ppm L. alba essential oil exhibited 100% mortality of treated mites at 48 hr exposure by fumigant toxicity method.	*Oligonychus coffeae* Nietner in tea crops.	Deka et al. 2021
6.	*Schinus molle* and *Bulnesia sarmientoi*	*B. sarmientoi* Guaiol, Bulnesol.	*B. sarmientoi* caused the highest mortality rate.	*Rhipicephalus evertsi*, *Rhipicephalus appendiculatus* and *Rhipicephalus pulchelus* ticks	Luns et al. 2021
7.	*Eucalyptus globulus*	Eucalyptol, 1,8-cineole, menthol, menthone.	At 10% concentration of the EO, 97% acaricidal mortality.	*Rhipicephalus* (Boophilus) *annulatus* ticks.	Adenubi et al. 2021

...*Table 5.1 contd.*

...Table 5.1 contd.

S. N.	EO	Constituents	Remarks	PEST	References
	INSECTICIDAL PROPERTY				
1.	–	Citral + Castor oil + Water borne PU **Release control**		Melon fly (*Bactrocera cucurbitae* (Coquillett))	Zheng et al. 2021
2.	*Zuccagnia punctata (Zp)*	thuja-2,4(10)-diene, dihydroeugenol, alpha-pinene, and (-)-5,6-dehydrocamphor.	(ZpEO) + poly(ε-caprolactone) matrixes loaded with ZEO (ZEOP) prepd. by solvent casting.	*Triatoma infestans* (Chagas disease transmitter)	Lopez et al. 2021
3.	*Cymbopogon schoenanthus* (L.) spreng		Different concentrations of emulsion were used in phytosanitary treatments on the cotton crop in comparison with vulgarized binary insecticides (lambdacyhalothrin/profenofos 15/300 gha^{-1} and cypermethrin/acetamiprid 36/8 gha^{-1}).	red bugs (*Dysdercus voelkeri*, Schmidt)- in cotton culture	Nadio et al. 2021
4.	Mentha haplocalyx, eucalyptus EO, Menthae Haplocalycis Herba Cymbopogonis Citrari Deodorant, peppermint essence.	**Release control**	The beneficial effect of the invention is a slow release of volatile gases of mosquito repellent effect, extend the effective time, release the elegant fragrance and pleasant odour, the mosquito repellent while achieving air-refreshing effects. In addition, the invented preparation. of the aromatic bead sustained-released mosquito insecticide, the formulation is stable, improves the stability of volatile oil, effectively preventing the deterioration of the crystal ball, extending the effective time, and does not contain any chemical additives, safe and nontoxic, easy to use and store, can be widely applied to vehicles, rooms, wardrobes, toilets, and other places, in order to reach the fresh air, mosquito repellent effect.	mosquito repelling crystal ball. Mentha haplocalyx 0.5%–1%, eucalyptus essential oil 0.2% –1 Menthae Haplocalycis Herba Cymbopogonis Citrari Deodorant 0.1–0.5%, 0.4–1%, peppermint essence: 0.5–1%, eucalyptus essential oil 0.2–0.5%, solubilizing agent 1.6–2%, alcohol 1.6–2%, the rest is pure water.	Yin 2021

S. N.	EO	Constituents	Remarks	PEST	References
5.	Lemon grass oil, eucalyptus oil, citronella oil, camphor oil, cedar wood oil.	lemon grass oil, eucalyptus oil, citronella oil, camphor oil, cedar wood oil + Hexane extension of five Tree seed oils Jatropha oil, Neem oil, Annona oil, Pongamia oil, Mahua oil.	The mortality percent of BPH nymphs at a dose of 1.0% concentration of different seed extensions showed significant insecticidal activity. The mortality percentage was evaluated with different non-edible and essential oils of five solvent extensions were, viz. hexane extract. Tree seed oils neem oil showed significant insecticidal activity 84.00% followed by pongamia oil 71.25% jatropha oil 66.00%, Mahua oil 60.40%, Annona oil 58.25% compared to control 5.60%. In essential Eucalyptus oil 84.40%, Lemon grass oil 74.40%, cedar wood oils 67.40%, citronella oils 64.40%, camphor oils 56.40% control 3.60%.	3rd instar nymphs of brown planthopper *Nilaparvata lugens.* **Dipping method** The mortality percent of BPH nymphs at a dose of 1.0% concn. of different seed exts. showed significant insecticidal activity.	Sainath et al. 2020
6.	*Cymbopogon citratus* Stapf, *Ocimum gratissimum L.*	thymol and p-cymene (O. gratissimum) myrcene, geranial and neral (C. citratus).	The EO of *O. gratissimum* at concns. of 2 and 5% showed insecticidal activity on all pests (biting-sucking and carpophagous), except the phyllophagous Syllepte derogata. *C. citratus*, at 1% concentration was particularly toxic to whiteflies (Bemisia tabaci), however, it favoured the action of beneficial insects, specifically black ants and ladybirds in the cotton plots, unlike the chemical product.	*Cotton Entomofauna.*	Kobenan et al. 2021

...Table 5.1 contd.

...Table 5.1 contd.

S. N.	EO	Constituents	Remarks	PEST	References
7.	*Cinnamomum zeylancium, Zingiber officinale, Cuminum cyminum*	Cinnamomum zeylancium, Zingiber officinale, Cuminum cyminum + Beauvericin nano gel 2%.	LC_{50}s of Beauvericin nano gel 2%: 38 and 41 *C. zeylancium*: 44 and 144 *C. cyminum*: 151 and 11 *Z. officinale*: 18 and 29 Against *S. Oryzae* and *S. granaries* respectively. **Post-harvest storage**	*Sitophilus oryzae* and *Sitophilus granarius* (Coleoptera: Curculionidae) harmful insects cause a damage to stored product seeds.	Sabbour 2020
8.	*Elsholtzia serrata*		The Elsholtzia plant volatile oil is used in spices, food additives, medicines, daily chemical products additives, pesticides, and antibacterial materials.		Ouyang et al. 2021 **PATENT**
9.	*Citrus paradisi*	linalool , α-terpineol, D-limonene, d-carvone, octanal, β-pinene, β-phellandrene, and α-pinene.	Crude grapefruit oil (GO1) contained 28 compounds. Concentrated grapefruit oil (GO2) contained 20 compounds. D-Limonene was the dominant constituent in both. Octanal was more toxic to workers as compared to the other four constituents.	*Solenopsis invicta.*	Zhang et al. 2021
10.	*Lavandula latifolia* L.	terpinen-4-ol, γ-terpinene, camphor, 1,8-cineole, lavandulol, linalool, β-elemene, β-pinene, 1,4-cineole, camphene, a terpineol, α-copaene.	Spike lavender EO had contact toxicity on *E. leucographa* upon topical treatment; EC_{50} = 0.37%.	*Euphoria leucographa* (Family: Scarabaeidae) infected in custard apple (*Annona muricata*) fruit.	Al-Ansari et al. 2021

S. N.	EO	Constituents	Remarks	PEST	References
11.	*Origanum marjorana, Matricaria chamomilla, Cuminum cyminum*, and *Basilicum ocimum.*	**Release control** Ethanol	encapsulation of four essential oils. EO moderate mortality at 10,000 mg/L. LC_{50} values of basil and marjoram are 992 mg/L & 3162 mg/L respectively. Nano emulsion had the highest mortality rate at 625 mg/L. LC_{50} of basil Nano Emulsion (NE) and marjoram NE in the laboratory. Strains are 45 mg/L and 188 mg/L resp.	*Aphis craccivora in* cowpea aphid.	Abdelaal et al. 2021
12.	*Artemisia annua* flowers	α-pinene 1,8-cineole, pinocarvone, β-selinene, ketone, artemisinin, and camphor	LC_{50} for oral and fumigant bioassays of *A. annua* essential oil; 1.204% W/V and 3.343 μL/L air, resp.	*Glyphodes pyloalis* Walker	Oftadeh et al. 2021
13.	Chestnut flower, *Azadirachta Indica* walnut	Chestnut – 20:80, *Azadirachta Indica* – 10–60, Walnut – 10–60	Essential oil extended from the plant, were used for repelling and preventing tea lesser leafhopper in tea garden. It is environment-friendly, difficult to generate pesticide resistance and convenient to use.	tea lesser leafhopper	Zhang et al. 2021 **PATENT**
14.	*Origanum vulgare* L., *Thymus vulgaris* L	*O. vulgare*: thymol, carvacrol and terpinen-4-ol. T. vulgaris: γ terpinene, thymol, and p-cymene.	*Ae. aegypti* larvae: LC_{50} values for *O. vulgare* and *T. vulgaris* 37.5 μg/mL and 38.9 μg/mL, respectively. For adult mosquitoes, the LC_{50} : 14.3 and 11.7 μg/mL, resp.	*Aedes aegypti* L.	De Oliveira et al. 2021
15.	*Ocimum basilicum, Mentha spicata,*	*O. basilicum*: Estragole, linalool. *M. spicata:* Cis-carvyl acetate, D-limonene.	*M. spicata*: at 6 hr and 24 hr of application at 5% induced 80 and 100% larval mortality, respectively. *O. basilicum*: at 24 hr treatment, at 2.5 and 5% concentrations, caused 80 and 100% larval mortality, respectively.	*Helicoverpa armigera* in cotton	Boulamtat et al. 2021

...Table 5.1 contd.

...Table 5.1 contd.

S. N.	EO	Constituents	Remarks	PEST	References
16.	Black pepper, eucalyptus, rosemary, tea tree.		EO Mortality Black pepper: 80% Tea tree: 80% Rosemary + TT: 98.33% Black pepper + TT: 98.33%	*Myzus persicae* (Aphididae: Hemiptera)	Ahmed et al. 2021
17.	*Mentha pulegium* and *Ferula gummosa*	**Release control**	When changed to nanoliposome, released and the stability of EOs were significantly affected. As compared with the crude EO of *M. pulegium*, its nanoliposome displayed enhancing fumigant toxicity against *T. castaneum.*	*Tribolium castaneum.*	Faraji et al. 2020
18.	*Zanthoxylum limonella*	α-fenchene, β-pinene, 9-octadeca none, D-limonene, p-mentha-1,5,8-triene, terpinen-4-ol, linalool, 1,8-cineole, α-thujene, decanal, γ-terpinene, α-phellandrene and linalyl propionate.	At 14 days, 10% of essential oil showed 100% mortality against the eggs of *T. castaneum.*	*Tribolium castaneum.*	Wanna and Satongrod 2020
19.	*Zanthoxylum alatum* (Z), *Myristica fragrans* (M), and *Bunium persicum* (B) (1:1:1 vol./vol./v) and their Nanoformulation (Ne-MBZ)	γ-terpinene, Linalool, and cuminaldehyde.	The pesticidal efficacy of nanoformulation (Ne-MBZ) against *C. chinensis* was significantly higher than individual.	*Callosobruchus chinensis* in beans.	Yadav et al. 2021
20.	*Piper aduncum*	α-cypermethrin, β-cyperm ethrin, fenpro pathrin, and γ-cyhalothrin.	The insecticides synergistic combination with EO presented a homogeneous response by topical and residual contact.	*Spodoptera frugiperda.*	Fazolin et al. 2016

S. N.	EO	Constituents	Remarks	PEST	References
21.	*Matricaria chamomilla* L., *Cuminum cyminum* L.	–	All EO were pathogenic on aphid.	*Aphis gossypii* Glover in cucumber.	Mousavi et al. 2020
22.	*Satureja intermedia*	γ-terpinene, p-cymene, carvacrol, and Thymol.	The EO possessed high fumigant toxicity against the four major stored-product insect pests.	Stored pests: *Oryzaephilus surinamensis*, *Tribolium castaneum*, *Rhyzopertha dominica*, and *Trogoderma granarium* Everts.	Ebadollahi and Setzer 2020
ANTIFUNGAL					
1.	*Ocimum basilicum* L.	Linalool	The basil essential oil and linalool; *F. oxysporum* (median inhibitory concentration (IC_{50}) = 13.18 and 12.60 μg/mL) respectively. Penicillium spp. (IC_{50} = 17.5 and 8.77 μg/mL) and C. gloeosporioides (IC_{50} = 12.91 and 9.15 μg/mL), respectively. The tomato and lettuce seeds' germination were not significantly affected by the essential oils.	*Colletotrichum gloeosporioides*, *Penicillium* spp. and *Fusarium oxysporum*.	Torre et al. 2021
2.	*Lavandula latifolia* L.	γ-terpinene camphor 1,8-cineole, lavandulol, terpinen-4-ol, β-elemene, β-pinene, 1,4-cineole, camphene, linalool, α-terpineol, α-copaene	The minimum inhibitory concentration value 0.125 ± 0.05 μg/mL	*Trichophyton mentagrophytes*	Al-Ansari et al. 2021
3.	*Citrus aurantium* ssp. bergamia		At 2500 μL L^{-1} – reduction of sporulation and mycelial growth of the pathogen by 29.48 and 68.15%, resp.	*Alternaria solani* in the tomato	Hendges et al. 2020
4.	*Zanthoxylum alatum* (Z), *Bunium persicum* (B), *Myristica fragrans* (M), and (1:1:1 vol./vol./v) and their Nanoformulation (Ne-MBZ)	γ-terpinene, Linalool, and cuminaldehyde	The pesticidal efficacy of nanoformulation (Ne-MBZ) against *A. flavus* and AFB_1 was significantly higher than individual.	*Aspergillus flavus* and aflatoxin B_1	Yadav et al. 2021

...*Table 5.1 contd.*

...Table 5.1 contd.

S. N.	EO	Constituents	Remarks	PEST	References
5.	*Zingiber officinale* Roscoe	α-zingiberene Gerania	The EO inhibited aflatoxin production and controlled fungal growth.	*Aspergillus flavus* in stored maize grains.	Nerilo et al. 2020
6.	*Coriandrum sativum* L.	Thymol, 1-decanol, sedanolide, dodecanal, trans-trans-tetradec-2-enal, decanal, menthone, 2-decen-1-ol, and trans-2-dodecen-1-ol.	The EO displayed excellent antimicrobial activity against the fungus.	Candida albicans ATCC 10231.	Foudah et al. 2021
7.	*Micromeria fruticosa serpyllifolia*	–		*Epidermophyton floccosum*	Salameh et al. 2020
8.	*Litsea cubeba*	**Release control**	Complex based on β-cyclodextrin (β-CD) and Litsea cubeba EO (LCEO) was able to control the growth of the tested fungi.	*Penicillium italicum, Penicillium digitatum.* *Geotrichum citri-aurantii* from postharvest citrus.	Wang et al. 2021
9.	*Ficus exasperata*	α-thujopsene, 1,8-cineole, sabinene, β-patchoulene, α-pinene, and α-terpineol.	The EO displayed a significant zone of inhibition against C. albicans with MIC of 1.1 μg/mL.	*Candida albicans.*	Oladosu et al. 2009

Table 5.2 The names and constituents of essential oils having antimicrobial activity

S. N.	EO	Constituents	Remarks	Microbes	References
	ANTIBACTERIAL				
1.	*Coriandrum sativum* L.	1-decanol, decanal, thymol , sedanolide, menthone, 2-decen-1-ol, trans-tetradec-2-enal, trans-dodecanal, and trans-2-dodecen-1-ol.	The EO displayed excellent antimicrobial activity against all microorganisms, except *Escherichia coli.*	Klebsiella pneumonia NCTC 9633, Staphylococcus aureus ATCC 25923, Bacillus subtilis ATCC 11774, Escherichia coli ATCC 11229.	Foudah et al. 2021
2.	*Oreganum vulgare*	carvacrol, γ-terpinene, p-cymene, α-terpinene.	Oregano EO was an effective antimicrobial agent against *E. faecalis.*	*Enterococcus faecalis*	Krishnamachari et al. 2019
3.	*Ferula aucheri*	Germacrene B and β-caryophyllene-flowers; cis-dihydroagarofuran and δ-cadinene-fruit; δ-cadinene and gurjunene–roots.	The fruit and root oils were more potent against *E. coli* than gentamicin.	*Bacillus subtilis, Staphylococcus aureus, Klebsiella pneumonia, Escherichia coli, Shigella dysenteriae, Salmonella paratyphi.*	Koulaei et al. 2020
4.	*Micromeria fruticosa serpyllifolia*	–	The EO from the southern region displayed the highest potency against *Shigella sonnei.*	*Pseudomonas aeruginosa, Escherichia coli, Shigella sonnei, Proteus mirabilis, Staphylococcus aureus Klebsiella pneumoniae Enterococcus faecium.*	Salameh et al. 2020
5.	Lasia spinosa (L.) Thwaites	camphor, α-pinene and δ-3-carene.			Rahman et al. 2019
6.	(Abies alba L.)	β-pinene, ispbornyl acetate, α-pinene, camphene	Zone of inhibition ranges from 14.33 ± 0.58 mm to 16.67 ± 1.15 mm.	*E. coli, K. pneumonia, P. vulgaris, P.aeruginosa, B. aureus, S. aureus, L. monocitogenes, B. subtilis, C. albican.*	Marjanovic-Balaban et al. 2020
7.	Haematostaphis barteri Hook. f. (Anacardiaceae)	ar-curcumene, β-sesquiphellandrene, α-zingiberene, β-bisabolene, (E)-caryophyllene and fatty acids, oleic acid, palmitic acid and linoleic acid.	Min. Inhibition Concentration range of 78.1 to 1250 μg/mL.	*Staphylococcus epidermidis Pseudomonas aeruginosa*	Owolabi et al. 2020

Essential Oils as Food Preservatives

Food spoilage has continued to be a source of massive economic loss in different parts of the world. Hence, the need for the development of natural preservatives in the food industry cannot be overemphasized and natural products hold a great promise in this regard (Ogunnupebi et al. 2020). The preference for essential oils as preservatives was observed by in a recent review by (Falleh et al. 2020). The development of EO-based preservatives have continued to gain more ground, especially because of the public perception about the toxicity of synthetic molecules and their generally recognized as safe (GRAS) regulatory status. Approaches that are currently used include synergistic association of essential oils from different families/sources; nanotechnological applications and the impregnation of food packaging materials with EOs.

Synergistic Association of EOs

Several combinations of bioactive plant extracts and EOs are documented in literature. This association could either result into an antagonistic or synergistic effect. A synergy occurs when the antimicrobial activity of the mixture is higher than the sum of individual essential oils when applied alone. The observation of synergy implies that increased effectiveness is achieved as the effective concentration is reduced. Selected examples of synergistic association among Eos from different sources are shown in Table 5.3.

Table 5.3 EOs with synergistic association

S. N.	EO Types	Remark	References
1.	*Menthapiperita, Ormenismixta* and *Menthapulegium* essential oils.	Using the mixture design methodology, a synergy was observed in the *Menthapulegium/Menthapiperita* mixture. The optimal combination ratios were 54/46% (*E. coli*), 65/44% (*S. aureus*) and 55/44% (*C. tropicalis*)	Chraibi et al. 2021
2.	Binary, tertiary and quartenary combination of the EOs of *Origanum. Compactum* (O), *Thymus satureioides* (T), *Cinnamomum verum* (C) and *Allium. sativum* (G) were prepared and investigated	The tertiary combinations of EOs from T, C and O showed the best result against *S. enterica, S aureus, and E. coli,* increasing the shelf life of boneless chicken thighs (by 3 d).	Shankar et al. 2022
3.	EOs of *Menthapulegium, Lavandulastoechas* and *Laurusnobilis.*	CUPRAC, DPPH and ABTS antioxidant assays show that *Laurusnobilis* EO interacted synergistically with the other two EOs at a ratio of 2:1.	Yakoubi et al. 2021
4.	EO/EO combination and EO/Main Active constituent combination were examined for possible synergy. *EOs were Illicium Cymbopogon, citratusverum* and *Myristicafragrans, while the* main active constituents were geranial, trans-anethole, α-pinene.	Although Formulation (I. *verum* EO: geranial; 1:1) had the strongest synergistic effect, the combination of *the three EOs* demonstrated strong synergistic effect against *M. domestica* and was more active than cypermethrin.	Aungtikun et al. 2021

Application of Nanotechnology in EO Applications

Although very attractive, EOs and other lipid-based bioactives exhibit low water solubility and high chemical instability (Bitterling et al. 2021). Therefore, more research effort has been directed towards improving the application of EOs in food preservation. A quicker developing approach which solves this problem is the impregnation of water-dispersible nanoemulsions with essential oils. This has in no small way, improved important physical properties, antibacterial and antioxidant potential of EOs. Furthermore, owing to the volatility, rapid decomposition, and poor bioavailability of EOs in the lower part of the intestines, nano-composite systems are now being developed to aid EO-delivery into the gastrointestinal tract (Yostawonkul et al. 2021). Furthermore, pure essential oil components are being applied individually. An example is the recent encapsulation of linalool into the chitosan nano-composite which successfully achieved the intended slow release of the preservative in rice (Das et al. 2021). These and other interesting nanotechnological advances are shown in Table 5.4.

Table 5.4 Nanotechnologoical advances in EO applications

S. N.	Technology	Remark	References
1.	EOs and EOs (of varying concentrations) plus their **nanoemulsions** were used in different treatments targeting *Clostridium sporogenes* in meat products.	The prepared nanoemulsions were more active than individual EOs (*cinnamon, Tahiti lemon, cardamom, chinese pepper oregano, cinnamon, cardamom, and chinese pepper)* judging from the reduction of growth in *C. sporogenes.*	Pinelli et al. 2021
2.	A three component (cinnamon, ginger, and cardamom) essential oils-nano emulsion was prepared via the self-emulsification technique.	The combination –68% cinnamon, 10% ginger and 22% cardamom EOs produced the **nanoemulsions** with the best properties (smallest average particle size and size distribution and the highest zeta potential). This combination also showed highest antioxidant and the best antimicrobial activity.	Jafarizadeh-Malmiri et al. 2022
3.	Cellulose **nanofibers** (CNF) were prepared as a coating and were charged with ginger EO and citric acid for enhanced preservative potential.	The CNF edible coating alone did not really increase the shelf life of the chicken meat. However, its combination with ginger EO and citric acid controlled the chemical/ microbial spoilage and prolonged the shelf-life of the meat.	Khaledian et al. 2019
4.	EOs from carvacrol, clove, and cumin, in addition to their **nanoemulsions** as antibacterial agents against *L. monocytogenes* and *S. flexneri.*	Compared to the essential oils, the nanoemulsions generally reduced the bacterial count within a relatively short time. Carvacrol nanoemulsion (at a low concentration 0.78%) exhibited the highest natural preservative potential in talaga cheese.	Elsherif and Al-Shrief 2021

Table 5.4 contd....

...Table 5.4 contd.

5.	Grapefruit peel EO-synthesized nanoemulsion against food borne-pathogenic bacteria.	The introduction of essential oil into the prepared nanoemulsion enhanced led to improved bacteriostatic activity.	Ozogul et al. 2020
6.	Water-dispersible *litseacubeba*-EO synthesized **nanoemulsions** with high stability	The reduction of *litseacubeba* EO size to nanoscale (diameter = 100 nm) led to better solubility and long-term stability at 4°C and 25°C. Consequently, the antibacterial, antibiofilm and antioxidant potential of *litseacubeba* EO nanoemulsion became stronger than the *litseacubeba* EO itself.	Wang et al. 2021
7.	Cinnamon EO **nanoemulsions** by high-pressure homogenization.	Although, gram-negative bacteria were more sensitive, cinnamon EO nanoemulsions exhibited good activity against both gram-positive and gram-negative bacterias. Its' highest bacteriostatic was against *Aspergillus niger* (12.5 μL/L MIC).	Liu et al. 2021
8.	Preparation of **nanoparticles** (with cinnamon oil) by the hot homogenization technique and their micro-particulate encapsulation via the alginate-calcium chloride system.	*Nano/Microstructured hybrid composite system showed excellent activity against important food borne pathogens (P. aeruginosa, E. coli, S. typhi, S.enterica, V. cholera, and S. aureus). It also achieved the intended intestinal target delivery and exhibited good gastric resistance and sustained the release in intestinal fluids.*	Yostawonkul et al. 2021
9.	Novel linalool/chitosan **nanocomposite** controlled release delivery system for the protection of stored rice against fungal infestation and lipid peroxidation.	The delivery of the preservative was successful with CN-linalool, inhibiting fungal and AFB1 contamination in rice; it also hindered lipid peroxidation and maintained sensory attributes in rice.	Das et al. 2021
10.	Alginate-based edible coatings (impregnated with *Trachyspermumammi* essential oil) were used as an antibacterial agent.	Alginate coating which was loaded with nanoemulsion gave better anti-listerial activity. It almost prevented the growth of *Listeria monocytogenes* (a food borne pathogen) in turkey fillets > 12 d (cold storage).	Kazemeini et al. 2021

Packaging Materials

The fact that oxidizing agents and several food-borne pathogenic organisms continue to threaten food safety has led to the development of different preservative technologies, including active packaging. This kind of packaging is usually impregnated with antimicrobial or/and antioxidants which are slowly released to extend the shelf life of the packaged product. Recently, the demand for safe and ecofriendly packaging materials has continued to be on the increase. In order to meet this demand, essential oils, again, because of their broad antimicrobial spectrum coupled with their GRAS status are widely investigated. Interesting advances in this area include the recent application of essential oils in the production of biodegradable food packaging (Souza et al. 2021). This and other applications of essential oils in active packaging are contained in Table 5.5.

Table 5.5 Application of EOs in active packaging

S. N.	Packaging Material	Remark	References
1.	Thermoplastic starch films (containing nanocellulose-stabilized Pickering emulsions of different essential oils: Ho wood, cardamom, and cinnamon) were prepared by the casting method and examined for their suitability as packaging materials.	Ho wood films demonstrated intense chemical interactions, better mechanical resistance and is therefore the most promising emulsion the preparation of biodegradable active packaging.	Souza et al. 2021
2.	Direct application of active EOs (in active packaging) to inhibit the growth of *L. grayi* on thawed frozen vegetable products.	Vapors of cinnamon and oregano EOs were effective in inhibiting the growth of *L. monocytogenes* and *L. gray in vitro*/cinnamon EO vapor inside the packaging inhibited *Listeria* growth on vegetables.	Tao et al. 2021
3.	Twenty-two (antioxidant) active packaging LDPE films were prepared using two essential oils and seven vegetable oils.	LDPE film/Flaxseed oil (50 μm) extended the shelf-life of fresh meat by 22%.	Wrona et al 2021.
4.	Modified Atmosphere Packaging (MAP) with thyme, clove, and cinnamon essential oil EO vapors for the preservation of two varieties of mango fruit.	Anthracnose, and the loss of fruit firmness were significantly reduced by MAP with thyme oil vapor treatment ($p < 0.05$) The same atmosphere extended the shelf life of cv. Banganapalli to 26 ± 2 d and cv. Totapuri to 23 ± 2 d.	Anand et al. 2021
5.	Bioactive edible films were prepared from chitosan by casting and electrospinning techniques and targeted at active food packaging. Varying quantities of Zataria multiflora and cinnamon essential oils were incorporated.	The bilayer film were generally active against *A. flavus* and the introduction of essential oil blends influenced the antioxidant potential of the materials. The properties of the film (transmittance and hydrophobicity) were also influenced by the addition of essential oils.	Ebrahimzadeh et al. 2021
6.	Pectin edible coating containing different mixtures of *Thymus vulgaris EO* and *Thymbraspicata* EO for the preservation of sliced bolognas.	EO treatment of the edible pectin reduced the mold counts, lactic acid bacteria, and controlled the growth of *Salmonella typhimurium*. It also extended the bolognas' shelf-life.	Gedikoğlu, 2022

Limitations in Usage of Essential Oils

Despite the attractiveness of essential oils in different applications, some limitations to their applicability exist. Essential oils can quickly depreciate in quality if not properly stored, because of external factures such as heat, light and oxygen (Bitterling et al. 2021). This was recently demonstrated by Zeljković et al. (2021) who proved that EOs depreciate in quality and potency during storage. Their work showed that the antioxidant activity of thyme and spearmint EOs decreased up to 25–70% within 25 wk of storage. Hence, proper quality control measures must be taken while incorporating EOs into various preservative technologies.

Another limitation of EOs which is being tackled by various research efforts is their solubility. The solubility challenge also affects the bioavailability of EOs in the human body (Yostawonkul et al. 2021). The high lipophilicity of EO constituents make them insoluble in water, thus limiting their applicability in certain foods and drinks (Sampaio et al. 2022).

The strong flavor of EOs also constitute a challenge to their application in food production as the EO's odor tampers with the organoleptic properties of the food (Sampaio et al. 2022). This limits the percentage of EOs which can be applied in some cases, thus reducing the antioxidant or antimicrobial effect which the EO was originally intended to contribute.

Conclusion

EOs has been found application in various fields including agriculture. They have great potential to replace or complement the synthetic agricultural products used in pre- and post-harvest pest control. They can be used alone or in combination with the synthetic products to reduce their toxicity. They can kill various pests (bacteria, viruses, and fungi) as well as nematodes and mites. The use of nanoemulsions as a carrier solves its solubility challenge. At the same time, their encapsulations in different edible polymers proffer solution to their release control. Therefore, EOs may constitute the most patronized agricultural pesticides in the next decade.

References

Abdelaal, K., M. Essawy, A. Quraytam, F. Abdallah, H. Mostafa, K. Shoueir, et al. 2021. Toxicity of essential oils nanoemulsion against *Aphis craccivora* and their inhibitory activity on insect enzymes. Processes. 9(4): 624.

Adenubi, O.T., A.O. Abolaji, T. Salihu, F.A. Akande and H. Lawal. 2021. Chemical composition and acaricidal activity of *Eucalyptus globulus* essential oil against the vector of tropical *Bovine piroplasmosis*, *Rhipicephalus* (Boophilus) *annulatus*. Exp. Appl. Acarol. 83(2): 301–312.

Ahmed, Q., M. Agarwal, R. Al-Obaidi, P. Wang and Y. Ren. 2021. Evaluation of aphicidal effect of essential oils and their synergistic effect against *Myzus persicae* (Sulzer) (Hemiptera: Aphididae). Molecules. 26(10): 3055.

Al-Ansari, M.M., A.M.I. Andeejani, E. Alnahmi, R.H. AlMalki, A. Masood, P. Vijayaraghavan, et al. 2021. Insecticidal, antimicrobial and antioxidant activities of essential oil from *Lavandula latifolia* L. and its deterrent effects on *Euphoria leucographa*. Ind. Crops. and Prod. 170: 113740.

Al-Saleh, I.A. 1994. Pesticides: a review article. J. Environ. Pathol. Toxicol. Oncol. 13(3): 151–161. PMID: 7722882.

Anand, B.P., B.N. Reshma, S.S. Periyar and S.E. Rotimi. 2021. Use of modified atmosphere packaging combined with essential oils for prolonging post-harvest shelf life of mango (cv. Banganapalli and cv. Totapuri), LWT. 148: 111662. ISSN 0023-6438, https://doi.org/10.1016/j.lwt.2021.111662.

Arabaci, T., S. Çelenk, T. Özcan, E. Martin, T. Yazici and M. Açar, et al. 2020. Homoploid Hybrids of *Origanum* (Lamiaceae) in Turkey: Morphological and Molecular Evidence for a New Hybrid. Plant Biol. 155(3): 470-482. https://doi.org/10.1080/11263504.2020.1762777.

Aungtikun, J., M. Soonwera and S. Sittichok. 2021. Insecticidal synergy of essential oils from *Cymbopogon citratus* (Stapf.), *Myristica fragrans* (Houtt.), and *Illicium verum* Hook. f. and their major active constituents. Ind. Crops Prod. 164: 113386. https://doi.org/10.1016/j.indcrop.2021.113386.

Bitterling, H., P. Lorenz, W. Vetter, D.R. Kammerer and F.C. Stintzing. 2021. Storage-related changes of terpene constituents in caraway (*Carum carvi* L.) under real-time storage conditions. Ind. Crops Prod. 170: 113782. https://doi.org/10.1016/j.indcrop.2021.113782.

Boulamtat, R., A. Mesfioui, K. El-Fakhouri, A. Oubayoucef, A. Sabraoui, A. Aasfar, et al. 2021. Chemical composition, and insecticidal activities of four plant essential oils from Morocco against larvae of *Helicoverpa armigera* (Hub.) under field and laboratory conditions. Crop Prot. 144: 105607.

Chraibi, M., M. Fadil, A. Farah, S. Lebrazi and K. Fikri-Benbrahim. 2021. Antimicrobial combined action of *Mentha pulegium*, *Ormenis mixta* and *Mentha piperita* essential oils against *S. aureus*, *E. coli* and *C. tropicalis*: Application of mixture design methodology. LWT. 145: 111352.doi:10.1016/j.lwt.2021.111352.

de Oliveira, A.A., L.P. Franca, A.S. Ramos, J.L.P. Ferreira, A.C.B. Maria, K.M.T. Oliveira, et al. 2021. Larvicidal, adulticidal and repellent activities against *Aedes aegypti* L. of two commonly used spices, *Origanum vulgare* L. and *Thymus vulgaris* L. S. Afr. J. Bot. 140: 17–24.

Das, S., V.K. Singh, A.K. Chaudhari, A.K. Dwivedy and N.K. Dubey. 2021. Fabrication, physico-chemical characterization, and bioactivity evaluation of chitosan-linalool composite nano-matrix as innovative controlled release delivery system for food preservation. Int. J. Biol. Macromol. 188: 751–763. doi: 10.1016/j.ijbiomac.2021.08.045.

Deka, B., A.K. Pandey, A. Babu, C. Baruah and S. Sarkar. 2021. Acaricidal and ovicidal properties of *Lippia alba* essential oil and its chemical constituents against red spider mite, *Oligonychus coffeae* Nietner (Acari: Tetranychidae) infesting tea crops. Arch. Phytopathol. Plant Prot. 54(19-20): 1738–1752.

Duque, L.S., P. Marchesini, C. Monteiro, G.A. Gomes, T.H. Soares Rodrigues, D.M. Mesquita, et al. 2021. Acaricidal activity of the essential oils from *Leptospermum scoparium*, *Origanum vulgare* and *Litsea cubeba* on *Rhipicephalus microplus*: influence of the solvents and search for fractions with higher bioactivity. Vet. Parasitol. 300: 109606.

Ebadollahi, A. and W.N Setzer. 2020. Evaluation of the toxicity of *Satureja intermedia* C. A. Mey essential oil to storage and greenhouse insect pests and a predator ladybird. Foods. 9(6): 712.

Ebrahimzadeh, S., M.R. Bari, H. Hamishehkar, H.S. Kafil and L.-T. Lim. 2021. Essential oils-loaded electrospun chitosan-poly(vinyl alcohol) nonwovens laminated on chitosan film as bilayer bioactive edible films. LWT. 144: 111217. doi:10.1016/j.lwt.2021.111217.

Elsherif, W.M. and L.M. Talaat Al Shrief. 2021. Effects of three essential oils and their nano-emulsions on *Listeria monocytogenes* and *Shigella flexneri* in Egyptian Talaga cheese. Int. J. Food Microbiol. 2(355): 109334. doi: 10.1016/j.ijfoodmicro.2021.109334.

Falleh, H., J.M. Ben, M. Saada and R. Ksouri. 2020. Essential oils: a promising eco-friendly food preservative. Food Chem. 2020 Nov. 15, 330: 127268. doi: 10.1016/j.foodchem.2020.127268. Epub 2020 Jun 8. PMID: 32540519.

Faraji, Z., J. Shakarami, J. Varshosaz and S. Jafari. 2020. Encapsulation of essential oils of *Mentha pulegium* and *Ferula gummosa* using nanoliposome technology as a safe botanical pesticide. J. Appl. Biotechnol. Rep. 7(4): 237–242.

Fazolin, M., J.L.V. Estrela, A.F.M. Medeiros, I.M. da Silva, L.P. Gomes and M.S.F. Silva. 2016. Synergistic potential of dillapiole-rich essential oil with synthetic pyrethroid insecticides against fall armyworm. Ciencia Rural. 46(3): 382–388.

Foudah, A.I., M.H. Alqarni, A. Alam, M. Ayman Salkini, E.O. Ibnouf Ahmed and H.S. Yusufoglu. 2021. Evaluation of the composition and *in vitro* antimicrobial, antioxidant, and anti-inflammatory activities of Cilantro (*Coriandrum sativum* L. leaves) cultivated in Saudi Arabia (Al-Kharj). Saudi J. Biol. Sci. 28(6): 3461–3468.

Gedikoğlu, Ayça. 2022. The effect of *Thymus vulgaris* and *Thymbra spicata* essential oils and/or extracts in pectin edible coating on the preservation of sliced bolognas. Meat. Sci. 184: 108697. https://doi.org/10.1016/j.meatsci.2021.108697.

Hendges, C., J.R. Stangarlin, M. de Holanda Nozaki, E. Lorenzetti and O.J. Kuhn. 2020. Atividade antifúngica contra Alternaria solani e controle da pinta preta do tomateiro por óleo essencial de bergamota. Semina: Ciências Agrárias. 41(5supl1): 1861–1874.

Jafarizadeh-Malmiri, H., N. Anarjan and A. Berenjian. 2022. Developing three-component ginger-cinnamon-cardamom composite essential oil nanoemulsion as natural food preservatives. Environ. Res. 204, Part B: 112133. https://doi.org/10.1016/j.envres.2021.112133.

Kazemeini, H., A. Azizian and H. Adib. 2021. Inhibition of *Listeria monocytogenes* growth in turkey fillets by alginate edible coating with *Trachyspermum ammi* essential oil nano-emulsion. Int. J. Food Microbiol. 344: 109104. doi: 10.1016/j.ijfoodmicro.2021.109104.

Khaledian, Y., M. Pajohi-Alamoti and B. Bazargani-Gilani. 2019. Development of cellulose nanofibers coating incorporated with ginger essential oil and citric acid to extend the shelf life of ready-to-cook barbecue chicken. J. Food Process. Preserv. 43(10): e14114. doi:10.1111/jfpp.14114.

Kobenan, K.C., K.K.N. Bini, M. Kouakou, I.S. Kouadio, G. Zengin, G.E.C. Ochou, et al., 2021. Chemical composition and spectrum of insecticidal activity of the essential oils of *Ocimum gratissimum* L. and *Cymbopogon citratus* Stapf on the main insects of the cotton entomofauna in Côte d'Ivoire. Chem Biodivers. 18(11): e2100497. doi: 10.1002/cbdv.202100497. .

Koulaei, S.A., A. Hadjiakhoondi, M.R. Delnavazi, Z. Tofighi, Y. Ajani, F. Kiashi, et al. 2020. Chemical composition and biological activity of *Ferula aucheri* essential oil. Res. J. Pharmacogn. 7(2): 21–31.

Krishnamachari, J., P. Ajitha, S. Raghu and V.T. Kavalipurapu. 2019. Chemical constituent, minimal inhibitory concentration, and antimicrobial efficiency of essential oil from *Oreganum vulgare* against *Enterococcus faecalis*: an *in vitro* study. J. Conserv. Dent. 22(6): 538–543.

Li, S., F. Zhao, S. Li and G. Liu. 2021. A kind of turmeric volatile oil and application. China patent # CN 113115799 A 20210716.

Liu, X., L. Chen, Y. Kang, D. He, B. Yang and K. Wu. 2021. Cinnamon essential oil nanoemulsions by high-pressure homogenization: Formulation, stability, and antimicrobial activity. LWT. 147: 111660. doi:10.1016/j.lwt.2021.111660.

Lopez, S., A. Tapia, J. Zygadlo, R. Stariolo, G.A. Abraham and P.R. Cortez Tornello. 2021. *Zuccagnia punctata* Cav. essential oil into poly(ε-caprolactone) matrices as a sustainable and environmentally friendly strategy biorepellent against *Triatoma infestans* (Klug) (Hemiptera, Reduviidae). Molecules. 26(13): 4056.

Luns, D.A.R., R. Martins, S. Pombal, J.M.L. Rodilla, N.W. Githaka and J.I.S. Vaz. 2021. Effect of essential oils against acaricide-susceptible and acaricide-resistant *Rhipicephalus* ticks. Exp. Appl. Acarol. 83(4): 597–608.

Marjanovic-Balaban, Z., V.G. Cvjetkovic, L. Stanojevic, J. Stanojevic, L. Nikolic and B. Danilovic. 2020. Quality testing of industrially produced essential oil of fir (*Abies alba* L.) from the republic of Srpska. J. Essent. Oil Bear. Plants. 23(3): 503–513.

Mousavi, M., Y. Ghosta and N. Maroofpour. 2020. Insecticidal activity and sublethal effects of *Beauveria bassiana* (Bals.-Criv.) Vuill. isolates and essential oils against *Aphis gossypii* Glover, 1877 (Hemiptera: Aphididae). Acta Agriculturae Slovenica. 115(2): 463–472.

Nadio, N.A., E.M. Bokobana, K.P. Akantetou, P. Tozoou, W. Poutouli, K. Koba, et al. 2021. Effectiveness of organic insecticides based on *Cymbopogon schoenanthus* (L.) spreng essential oil against red bunches (*Dysdercus voelkeri*, Schmidt) in cotton culture in Togo. Afr. J. Food Agric. Nutr. Dev. 21(3): 17727–17740.

Nerilo, S.B., J.C.Z. Romoli, L.P. Nakasugi, N.S. Zampieri, S.A.G. Mossini, G.H.O. Rocha, et al. 2020. Antifungal activity and inhibition of aflatoxins production by *Zingiber officinale* Roscoe essential oil against *Aspergillus flavus* in stored maize grains. Ciencia Rural. 50(6): e20190779.

Oftadeh, M., J.J. Sendi, A. Ebadollahi, Setzer, N. William and P. Krutmuang. 2021. Mulberry protection through flowering-stage essential oil of *Artemisia annua* against the lesser mulberry pyralid, *Glyphodes pyloalis* walker. Foods. 10(2): 210.

Ogunnupebi, T.A., A.P. Oluyori, A.O. Dada, O.S. Oladeji, A.A. Inyinbor and G.O. Egharevba. 2020. promising natural products in crop protection and food preservation: basis, advances, and future prospects. Int. J. Agron. 1–28.

Oladosu, I.A., M.F. Zubair, M.S. Ali and N.O. Olawore. 2009. Anticandidal activity of volatile compounds from the root bark of *Ficus exasperata* vahl-holl (Moraceae). J. Essent. Oil Bear. Plants. 12(5): 562–568.

Owolabi, M.S., A.L. Ogundajo, N.S. Dosoky and W.N. Setzer. 2020. Chemical composition and antimicrobial potential of essential oils of leaf and stem bark of *Haematostaphis barteri* Hook. f. (Anacardiaceae). J. Essent. Oil Bear. Plants. 23(3): 583–593.

Ouyang, W., J. Lium and C. Gan. 2021. Preparation method and application of high-quality and efficiency Elsholtzia plant volatile oil. China patent # CN 113416609 A 20210921.

Oyagbemi, T.O., A. Ashafa and J.O. Adejinmi. 2019. Evaluation of acaricidal activity of leaves of *Ficus exasperata* on dog ticks: *Rhipicephalus sanguineus*. J. Pharmacogn. Phytochem. 8(3): 1148–1153.

Özogul, Y., F. Özogul and P. Kulawik. 2021. The antimicrobial effect of grapefruit peel essential oil and its nanoemulsion on fish spoilage bacteria and food-borne pathogens. LWT. 136 (2): 110362. https://doi.org/10.1016/j.lwt.2020.110362.

Pinelli, J.J., M. Helena de Abreu, H., A.S. Guimarães, S.R. Isidoro, M.C. Gonçalves, T.S. Junqueira de Moraes, et al. 2021. Essential oil nanoemulsions for the control of *Clostridium sporogenes* in cooked meat product: An alternative? LWT. 143: 111123. doi:10.1016/j.lwt.2021.111123.

Prakash, B., A. Kujur, A. Yadav, A. Kumar, P.P. Singh and N. Dubey. 2018. Nanoencapsulation: An efficient technology to boost the antimicrobial potential of plant essential oils in food system. Food Control. 89: 1–11.

Rahman, A., S.A. Siddiqui, F. Oke-Altuntas, S. Okay, F. Gul and I. Demirtas. 2019. Phenolic profile, essential oil composition and bioactivity of *Lasia spinosa* (L.) thwaites. Braz. Arch. Biol. Technol. 62: e19170757.

Reichling, J., C. Koch, E. Stahl-Biskup, C. Sojka and P. Schnitzler. 2005. Virucidal activity of a beta-triketone-rich essential oil of *Leptospermum scoparium* (manuka oil) against HSV-1 and HSV-2 in cell culture. Planta. Medica. 71(12): 1123– 1127.

Sabbour, M.M.A. 2020. Insecticidal effect of three essential oils and Beauvericin nano gel on *Sitophilus oryzae* and *Sitophilus granarius* (Coleoptera: Curculionidae). J. Biopestic. 13(2): 127–134.

Sainath, G., V.K. Dubey and B.J. Rani. 2020. Insecticidal activity of different tree seed edible seed oils and essential oils against brown planthopper, *Nilaparvata lugens*, (Stal.) (Delphacidae: Homoptera). Int. J. Curr. Microbiol. Appl. Sci. 10: 729–735.

Salameh, N., N. Shraim, N. Jaradat, M. El Masri, L. Adwan, S. K'aibni, et al. 2020. Screening of antioxidant and antimicrobial activity of *Micromeria fruticosa serpyllifolia* volatile oils: a comparative study of plants collected from different regions of West Bank, Palestine. BioMed Res. Int. 4851879.

Sampaio, C.I., A.I. Bourbon, C. Gonçalves, L.M. Pastrana, A.M. Dias and M.A. Cerqueira. 2022. Low energy nanoemulsions as carriers of thyme and lemon balm essential oils. LWT. 154: 112748. https://doi.org/10.1016/j.lwt.2021.112748.

Shankar, S., S. Karboune, S. Salmieri and M. Lacroix. 2022. Development of antimicrobial formulation based on essential oils and gamma irradiation to increase the shelf life of boneless chicken thighs. Radiat. Phys. Chem. 19: 109893. https://doi.org/10.1016/j.radphyschem.2021.109893.

Souza, A.G., R.R. Ferreira, L.C. Paula, S.K. Mitra and D.S. Rosa. 2021. Starch-based films enriched with nanocellulose-stabilized Pickering emulsions containing different essential oils for possible applications in food packaging. Food Packag. Shelf Life. 27: 100615. doi:10.1016/j.fpsl.2020.100615.

Tao, R., J. Sedman and A. Ismail. 2021. Antimicrobial activity of various essential oils and their application in active packaging of frozen vegetable products. Food Chem. 360: 129956. doi:10.1016/j.foodchem.2021.129956.

Torre, R., E.A.D. Pereira, R.V. Nascimento, T.F. Guedes, P.R.S. Faria, M.S. Alves, et al. 2021. Agroecological approach to seed protection using basil essential oil. Ind. Crops Prod. 171: 113932.

Wanna, R. and B. Satongrod. 2020. Potential effects of essential oil from *Zanthoxylum limonella* seeds against *Tribolium castaneum* (Coleoptera: Tenebrionidae). Aust. J. Crop Sci. 14(12): 1920–1925. doi: 10.21475/ajcs.20.14.12.2770.

Wang, Y., C. Cen, J. Chen, C. Zhou and L. Fu. 2021. Nano-emulsification improves physical properties and bioactivities of *Litsea cubeba* essential oil. LWT. 137: 110361. https://doi.org/10.1016/j.lwt.2020.110361.

Weiss, B., S. Amler and R.W. Amler. 2004. Pesticides. Pediatrics 4: 1030–1036. PMID: 15060196.

Wrona, M., F. Silva, J. Salafranca, C. Nerín, M.J. Alfonso and M.A. Caballero. 2021. Design of new natural antioxidant active packaging: screening flowsheet from pure essential oils and vegetable oils to ex vivo testing in meat samples. Food Control, 120:107536

Yadav, A., A. Kumar, P.P. Singh and B. Prakash. 2021. Pesticidal efficacy, mode of action and safety limits profile of essential oils based nanoformulation against *Callosobruchus chinensis* and *Aspergillus flavus*. Pestic. Biochem. Physiol. 175: 104813.

Yakoubi, R., S. Megateli, T.H. Sadok, C. Bensouici and E. Bağci. 2021. A synergistic interactions of Algerian essential oils of *Laurus nobilis* L., *Lavandula stoechas* L. and *Mentha pulegium* L. on anticholinesterase and antioxidant activities. Biocatal. Agric. Biotechnol. 31: 101891. https://doi.org/10.1016/j.bcab.2020.101891.

Yostawonkul, J., N. Nittayasut, A. Phasuk, R. Junchay, S. Boonrungsiman, S. Temisak, et al. 2021. Nano/microstructured hybrid composite particles containing cinnamon oil as an antibiotic alternative against food-borne pathogens. J. Food Eng. 290: 110209. https://doi.org/10.1016/j.jfoodeng.2020.110209.

Yin, T. 2021. A kind of repelling mosquito crystal ball and preparation method thereof. China patent # CN 113475538 A 20211008.

Zeljković, S.C., K. Smékalová, K. Kaffková and N. Štefelová. 2021. Influence of post-harvesting period on quality of thyme and spearmint essential oils. J. Appl. Res. Med. Aromat. Plants. 25: 100335. https://doi.org/10.1016/j.jarmap.2021.100335.

Zhao, J., Y. Li, J. Xu, T. Liu, M. Zou and Q. Xue. 2021. Acaricide containing plant essential oil and using method and application thereof. China patent # CN 113615717 A 20211109.

Zhang, Y., H. Chen and K. Pan. 2017. Nanoencapsulation of food antimicrobial agents and essential oils. pp. 183–221. *In*: S.M. Jafari (ed.). Nanoencapsulation of Food Bioactive Ingredients. Academic Press: Cambridge, MA, USA.

Zhang, N., Y., Liao, L. Xie, Z. Zhang and W. Hu. 2021. Using essential oils from *Citrus paradisi* as a fumigant for *Solenopsis invicta* workers and evaluating the oils effect on worker behavior. Environ. Sci. Pollut. Res. 28(42): 59665–59672.

Zhang, T., J. Bai, C. Ma, C. Zhou, M. Song, Y. Zhang, et al. 2021. Plant essential oil for repelling tea lesser leafhopper and preparation method thereof. China patent # CN 113243396 A 20210813.

Zheng, F., T. Li, H. Xu, P. Hu, R. Wang, Z. Zhang, et al. 2021. Long-lasting repellent activities of eco-friendly polyurethane system for controlled citral against melon fly. Crop Prot. 148: 105745.

6

Potential of Essential Oils as Mosquito Repellent

Sara Omer Swar[1], *Saba Mehnaz*[2*] *and Rao Zahid Abbas*[2*]

Introduction

Repellents are chemicals that inhibit insects from flying and sitting on the skin of animals and human beings. These chemical substances inhibit insects from causing abrasions on the skin (Kline et al. 2022, Choochote et al. 2007). Insect repellents function through forming suspension blocks between insects and the surface (Deka et al. 2021). Essential Oil (EO), that is a combination of complex volatile molecules are derived from a variety of plants. Essential oils exhibit their effects against a number of hematophagous arthropods (Pereira et al. 2021). Some of them are used in commercial repellent compositions (Li et al. 2021, Curtis et al. 1989). Numerous repellents were examined worldwide, and synthetic repellents were formed as a result. Synthetic repellents are now widely used to minimize the damage caused by different insects around the world in fruits and stored grains (Dethier 1956). Vectors that transmit parasites are these days controlled by insecticides. Many of which get resistance to these chemicals. Due to global warming, mosquitoes that spread yellow fever, malaria and dengue fever have been shifted towards temperate and higher altitude areas. These mosquitoes then harm human beings that are susceptible to various diseases. The global estimate shows that the world's population residing inside the possible transmission zone for malaria has been predicted to raise from 45 to 60% (Canelas et al. 2021, EPA 1997).

Synthetic compounds have proved useful to prevent humans from a number of diseases especially by the piercing parts of insects (mosquitos). The DEET (N, Ndiethyl-m-toluamide) is a synthetic compound which is not only effective on the skin, but also the most widely used compound for killing mosquitoes (Isman 2006). These compounds cause environmental and human health problems as it is

[1] Department of Food Technology, College of Agricultural Engineering Sciences, Salahaddin University, Kurdistan, Iraq.

[2] Department of Parasitology, Faculty of Veterinary Sciences, University of Agriculture, Faisalabad.

* Corresponding authors: sabamehnaz2012@gmail.com; raouaf@hotmail.com

used by 30% of the world's population. As a result, there is a rise in natural and environmentally safe repellents (Pitasawat et al. 2003). Some repellents derived from plants are better than, synthetic repellents. Essential oil repellents have a short-term effectiveness due to their volatility. Furthermore, synthetic repellents are usually highly effective and last for longer period of time than natural repellents (Fradin and Day 2002, Junkum et al. 2021).

Numerous plants and pesticides have been tested during the last 50 yr. Different oils like lemon, eucalyptus, and citronella oils have been approved by US Environmental Protection Agency as insect repellent ingredients. These oils have low toxicity to the skin. Hence these are widely used throughout the world. Natural pest management products are not much safer as compared to synthetic pest control agents (Sukumar et al. 1991). The natural pest management safety level has not yet been determined. Some have adverse side effects to medication (Katz et al. 2008). Research on most of them is inadequate and conflicting (Foster et al. 2005). Essential oils derived from these species like *Dacrydium franklini* and *Melaleuca bracteata* have been found to have promising qualities such as insect repellents, although they are often inappropriate to use due to skin sensitivity (Jacobson 1966). It was proposed that using these products to repel insects and minimize insects, chemicals should be applied to the specified area on the body (Trumble 2002, Elsayed and Hassabo 2021). This process helps to check if any hypersensitivity response could occur earlier treating the body. The focus of this chapter is on essential oils repellent activity and the synergistic effects produced by these mixtures on mosquitoes.

Essential Oils

The combination of a highly intricate volatile organic substance is known as essential Oils (EO). Within plants they act as secondary metabolites. These essential oils include hydrocarbons such as terpenes, sesquiterpenes. Some oxygenated compounds, i.e., alcohols, esters, ethers, aldehydes, ketones, lactones, phenols, and phenol ethers are also included in EO (Guenther 1972). These compounds are known due to their typical plant odors. There are more than three thousand essential oils known throughout the world (FAO 2018). However, 10% essential oils are commercially used in the various ingredients like makeup products, food items, and medicine industries. EO are considered nontoxic by US Food and Drug Administration Authority, and have no adverse effect (Arena et al. 2021). The composition of some scented plant species throughout the world can vary significantly (Zygadlo and Juiliani 2003). EO detected via different plant families have high repellency against insects as shown in Table 6.1. Major Genera (*Cymbopogon spp.*, *Eucalyptus spp.*, and *Ocimum spp.*) have been widely found among EO-producing plants. In tropical forest areas (the Bolivian Amazon) cymbopogon plants have been used from centuries to ward off insects like mosquitoes (Moore et al. 2007). Widely utilized insect repellents are produced by this genus and the variety of EO and their extracts are derived from these plants (Trongtokit et al. 2005). Efficacy of EO have been checked against various insects when tested in the laboratory contrary to *Anopheles arabiensis*, *Cymbopogon excavatus* commonly known as kachi grass showed 100% repellent effect for 2 hr. Its efficacy decreased to 59% when tested after 4 hr. The same plant extract was combined with 5% vanillin and tested against different types of mosquitoes, including the genus of Aedes, Cluex and Anopheles. It showed 100% protection for 6 hrs, as compared to 25% DEET (N,N-diethyl-3-methylbenzanmide) (Tawatsin et al. 2001). Field tests demonstrated full repellent effect for 11 hr against Anopheles mosquitoes by using the *Cymbopogon martini* oil (Ansari and Razdan 1994). *Cymbopogon citrus* is effective with paraffin (Oyedele et al. 1999).

Essential oils derived from the eucalyptus genus have been shown to have good repellent qualities. EO show a repellent effect for *Ixodes ricinus*, *Aedes albopictus*, and *Mansonia,* but show less repellency to Aedes aegypti and *C. pomonella* larvae (Jaenson et al. 2006, Toloza et al. 2008). However, EO have no effect on *Lasioderma serricorne* (Trongtokit et al. 2005). Traditional repellents have been made from the genus Ocimum spp. and used as oils. Suspensions made by adding paraffin suspension with *Ocimum gratissimum* and *Ocimum basilicum* showed a good repellent effect and protect from mosquito bites during cage conditions for about 8 hr (Gbolade et al. 2000). *Ocimum basilicum* was ineffective against the genus Lasioderma and Cydia (Hori 2003).

Essential oil as Metabolite

S Terpinoides like monoterpenes are common constituents of essential oils. They have been shown to have mosquito repellent properties (Jaenson et al. 2006). Beta caryophyllene known as the widely known sesquiterpene which works as a potent repellant against *Aedes aegypti* (Gillij et al. 2008). A linear diterpene alcohol such as phytol, displays a high repellent action against *Anopheles gambiae.* An amazing fact is that the repellent activities of various EO are linked with monoterpenoides as well as sesquiterpenes (Kiran and Devi 2007). Furthermore, the oxygen-containing chemicals (alcohol, beta citronellol, geraniol, and alpha pinene) were extracted through EO of DIANTHUS that have high tick repellent properties against ticks of genus Ixodes (Tunon et al. 2006). *Tribolium castaneum* was tested using metabolites obtained from *Baccharis salicifolia*, as well as some derivatives made by decreasing C=O and C=C bonds in germacrone and pulegone (Garcia et al. 2005). Aterpineol, pulegol, and germacrol, which include hydroxyls, are the most repelling chemicals. This indicates that the hydroxyl group has a positive impact on the *T. castaneum* repellent activity.

The repellent effect of more than 15 EO were tested with Argentine plants. Alcohols were found to be very efficient against *Pediculus humanus capitis.* The repellent effect of different Kenyan plants was tested against *A. gambiae* (Diptera). The Perillyl alcohol, cisverbenol, cis-carveol, geraniol, citronellal, perillaldehyde, caryophyllene oxide, carvacrol, four-isopropyl benzenemethanol, thymol, three-carene, and myrcene were found potent repellent compounds. In a number of mosquito repellent compounds, the OH (hydroxyl) group is associated with carbon. The activity against *Anopheles gambiae* was decreased by using different metabolites with the OH group associated with linalool, a-terpineol, and limonene (Wang et al. 2008). Terpenoids with two functional groups have been shown to be effective against mosquito repellents. According to the conceptual model, the positive end is very effective for the receptor interaction.

Table 6.1 Repellent effect of essential oils on mosquito

S. N.	Country	Animal	Animal Species	Plants used as Essential oil	References
1	India	Mosquito	*Anopheles annularis*	*Mentha piperita*	Ansari et al. 2000
2	India	Mosquito	*Anopheles culicifacies*	*Mentha piperita*	-do-
3	India	Mosquito	*Culex quinquefasciatus*	*Mentha piperita*	-do-
4	India	Mosquito	*Anpheles stephensi*	*Oscimum basilicum*	Prajapati et al. 2005
5	India	Mosquito	*Aedes Aegypti*	*Oscimum basilicum*	-do-
6	India	Mosquito	*Culex quinquefasciatus*	*Oscimum basilicum*	-do-
7	India	Mosquito	*Anpheles stephensi*	*Rosmarinus officinalis*	-do-
8	India	Mosquito	*Aedes Aegypti*	*Rosmarinus officinalis*	-do-
9	India	Mosquito	*Culex quinquefasciatus*	*Rosmarinus officinalis*	-do-
10	India	Mosquito	*Culex quinquefasciatus*	*Rosmarinus officinalis*	Pushpanathan et al. 2006
11	Thailand	Mosquito	*Aedes Aegypti*	*Zanthoxylum piperitum*	Choochote et al. 2007
12	Thailand	Mosquito	*Anopheles dirus*	*Curcuma longa L.*	Tawatsin et al. 2001
13	Thailand	Mosquito	*Culex quinquefasciatus*	*Cymbopogon winterianus*	-do-
14	Thailand	Mosquito	*Culex quinquefasciatus*	*Lamiaceae*	-do-
15	Thailand	Mosquito	*Anopheles dirus*	*Zanthoxylum limonella*	Trongtokit et al. 2005
16	Thailand	Mosquito	*Aedes Aegypti*	*Zanthoxylum limonella*	-do-
17	Thailand	Mosquito	*Culex quinquefasciatus*	*Zanthoxylum limonella*	-do-
18	Thailand	Mosquito	*Anopheles dirus*	*Pogostemon cablin*	-do-
19	Thailand	Mosquito	*Aedes Aegypti*	*Pogostemon cablin*	-do-
20	Thailand	Mosquito	*Culex quinquefasciatus*	*Pogostemon cablin*	-do-
21	Thailand	Mosquito	*Anopheles dirus*	*Syzygium cablin*	-do-
22	Thailand	Mosquito	*Aedes Aegypti*	*Syzygium cablin*	-do-
23	Turkey	Mosquito	*Culex pipiens*	*Pimpinella anisum*	Erler et al. 2006
24	Turkey	Mosquito	*Culex pipiens*	*Oscimum basilicum*	-do-
25	Turkey	Mosquito	*Culex pipiens*	*Eucalyptus camaldulensis*	-do-
26	Argentina	Mosquito	*Aedes Aegypti*	*Baccharis spartioides*	Gillij et al. 2008
27	Argentina	Mosquito	*Aedes Aegypti*	*Aloysia citriodora*	-do-

S. N.	Country	Animal	Animal Species	Plants used as Essential oil	References
28	Kenya	Mosquito	*Anopheles gambiae*	*Croton pseudopulchellus*	Odalo et al. 2005
29	Kenya	Mosquito	*Anopheles gambiae*	*Mkilua fragrans*	-do-
30	Kenya	Mosquito	*Anopheles gambiae*	*Endostemon tereticaulis*	-do-
31	Kenya	Mosquito	*Anopheles gambiae*	*Ocimum forskolei*	-do-
32	Kenya	Mosquito	*Anopheles gambiae*	*Ocimum fischeri*	-do-
33	Kenya	Mosquito	*Anopheles gambiae*	*Plectranthus longipes*	-do-
34	Kenya	Mosquito	*Anopheles gambiae*	*Conyza newii*	-do-
35	Kenya	Mosquito	*Anopheles gambiae*	*Tarchonanthus camphoratus*	Omolo et al. 2004
36	Kenya	Mosquito	*Anopheles gambiae*	*Tetradenia riparia*	-do-
37	Kenya	Mosquito	*Anopheles gambiae*	*Lippia javanica*	-do-
38	Kenya	Mosquito	*Anopheles gambiae*	*Lippia ukambensis*	-do-
39	Ethiopia	Mosquito	*Mansonia*	*Eucalyptus maculate citriodon*	Hadis et al. 2003
40	Nigeria	Mosquito	*Aedes Aegypti*	*Cymbopogon citratus*	Oyedele et al. 2002
41	Commercial Product	Mosquito	*Aedes Albopictus*	*Eucalyptus globulus*	Yang and Ma 2005
42	Sweden	Mosquito	*Aedes Aegypti*	*Dianthus caryophyllus*	Tunon et al. 2006

Essential Oils Synergistic Phenomena

The repulsive effect of EO is often attributed to specific molecules. Higher bioactivity among these metabolites takes place due to the synergistic case than the isolated components (Hummelbrunner and Isman 2001, Gillij et al. 2008). A researcher, Omolo et al. (2004) examined the repellant effects of essential oils and synthetic blends containing their principal ingredients. The activities of some of these mixes were significantly lower than those of the matching EO. This demonstrates the relevance of the composition in providing bioactivity to natural terpenoid mixtures. Minor elements also contribute to the repellant activity (Berenbaum 1985). It is considered that the green leaves present their defenses as a group of compounds rather than separate compounds. As a result, minor constituents in minute concentrations may act as synergists. This synergistic effect can also be seen in oil mixes. In 2006, researchers concluded that the repellent effect of a blend of essential oils from *Artemisia princeps* and *Cinnamomum camphora* against adults of *Sitophillus oryzae L.* and *Bruchus rugimanus* (Coleoptera) showed significantly higher repellent activity of individual oils. The synergistic effect of the primary components in the oils resulted in the EO's strong insect repellency and toxicity. Further research showed that the interactions between the EO showed improved repellent properties (Liu et al. 2006).

Essential Oil Repellent Effect

Essential oils show good results when they are used in a fresh form. Their beneficial effects are usually short-lived. Essential oils often work on mosquitoes when present in a vapor phase (Trongtokit et al. 2005). It works for short period period of time (Barnard, 2000). The volatile nature of EO has a significant impact on the rise in repellent activity of essential oils (Barnard 2000). The repellency length was extended using preparations that are formed by adding ointments, polymer mixes, microcapsules for the controlled release or creams. To enhance the repellent effect, *Zanthoxylum limonella* oil was micro coated with a glutaraldehyde cross-linked gelatin (Chang et al. 2006). The effectiveness of a mosquito repellent product containing *Cymbopogon citratus* oil on the external skin was analyzed during various preparations (Maji et al. 2007). First the efficacy decreases for the hydrophilic base followed by the emulsion and oleaginous bases (Oyedele et al. 2002).

To obtain the maximum repulsion effect of essential oils, paraffin, vanillin, salicylic acid, mustard and coconut oils have been mostly used as fixative agents (Oyedele et al. 2002). Vanillin is used to mix with essential oils and it significantly raises the protection situation of the EO when they are used against Aedes. It is also effective against other species of mosquitoes.

Fixative agents such as coconut oil, liquid paraffin, vanillin, salicylic acid, and mustard were used to boost the EO repellent's effectiveness. Vanillin is a substance that was once combined with EO (Choochote et al. 2007). When these substances were employed against *Aedes aegypti*, this chemical greatly extends the protection periods (PTs) of the EO (Khan et al. 1975). By adding 5% vanillin to 15% alcohol of *Eucalyptus globulus* proved to be effective against other mosquito species (Kamsuk et al. 2006). Researchers were able to extend the protection time against *A. albopictus*. It is notable that adding vanillin to *Zanthoxylum piperitum's* essential oil makes *Armigeres subalbatus*, a DEET-tolerant species, repellent (Kamsuk et al. 2006). However, the synthetic ingredients are cheap and natural active ingredients are relatively much more expensive.

Methods to Detect Repellent Activity

Synthetic ingredients have a lower market price as compared to natural active ingredients. The EO of *Z. piperitum* can be used to check the synergistic repellent effect of EOs. However, DEET is impracticable and, also ineffective. When enhancing the repellency of essential oils, numerous variables should be taken into account: New preparations, fixing ingredients, and the formation of complex repellents are all the possibilities for enhancing the strength as well as economic value of repellents formed from essential oils. Depending on the species, a number of approaches are used to evaluate the repellant activity. The arm-in-cage assay is the most used way to test EO in the Order Diptera. In history the first well-planned laboratory test on mosquito repellency, was reported in 1919 (Schreck 1977). EO was formally mixed in the solvent volume and uniformly administered to the arm of the volunteer (from the wrist to the elbow), and the solvent treatment was done on the other arm (Yang and Ma 2005). The product-treated arm and the control arm are placed in separate boxes containing a set number of blood-starved female mosquitoes. The number of mosquito bites in the given time were recorded. The repellent action of arthropods belonging to the Order Coleoptera, on the other hand, is typically measured using Petri dish experiments (Chaubey 2007). The two halves of filter paper are separated. A consistent amount of EO diluted in a volatile solvent, usually acetone is added at one end and the other half receives an equal quantity of solvent as a control. Both halves are rejoined, and the pre-determined number of insects (usually 20) are released on the paper's center. The number of arthropods on both halves is counted after many hours, and the percentage of the repellent effect was measured (Chaubey 2007, Wang et al. 2006).

The Y-tube olfactometer bioassay is used to assess the repellent effect for insects. These insects belong to the Coleoptera, Diptera, Thysanoptera, and Lepidoptera orders (Ndungu et al. 1999). It contains the crystal of Y-tube having a primary and secondary one that contains the repellents and the other which contains the control (Erler et al. 2006, Koschier and Sedy 2003, Katerinopoulos et al. 2005). The Flow of the air current is formed through the passage of air into the two arms of the Y-tube by a pump linked with the stem. The samples containing the EO and the control are placed on a piece of paper connected with the tube's arms. Arthropods are inserted in the tube through a hole in the middle (the place where the three tubes meet). The opening is covered with a rubber stopper. Then the pump is turned on. The number of insects in both tubes (treated and control) is tallied after seconds of exposure to determine the percentage of repellent effect (Erler et al. 2006). Another type of olfactorometer has been developed to assess the repellent activity, in which the moth larvae travels against the wind (Landolt et al. 1999). Due to the number of approaches used to assess repellency in literature, each author uses various settings, variations, and parameters, while quantifying the repellent activity, making comparisons between results from different studies around the world is difficult. Essential oils are classified in a repellency grade. However, the comparison of results can only be made by a positive or negative control.

Conclusion

Essential oils of plants and their specific metabolites have shown to be effective repellents against a number of species, including insects and other arthropods. Their extreme volatility, on the other hand, reduces the amount of time and can be used for protection from insects. As a result, when these natural chemicals are combined with fixative materials, they have a significant potential as commercial insect repellents, enhancing their efficiency and getting even greater activity than DEET, an effective mosquito repellent. Numerous oxygenated metabolites extracted from EO have shown the best repellent activity.

References

Ansari, M.A., P. Vasudevan, M. Tandon and R.K. Razdan. 2000. Larvicidal and mosquito repellent action of peppermint (Mentha piperita) oil. Bioresour. Technol. 71: 267–271.

Ansari, M.A. and R.K. Razdan. 1994. Repellent action of *Cymbopogon martini* martini Stapf. var. Sofia against mosquitoes. Indian J. Malariol. 31: 95–102.

Arena, M.E.., M.R. Alberto and E. Cartagena. 2021. Potential use of *Citrus* essential oils against acute respiratory syndrome caused by coronavirus. J. Essential Oil Res. 33: 330–341.

Barnard, D. 2000. Repellents and toxicants for personal protection. *In:* World Health Organization, Department of Control, Prevention and Eradication, Programme on Communicable Diseases, WHO Pesticide Evaluation Scheme (WHOPES), WHO/CDS/WHOPES/GCDPP/2000.5, WHO, Geneva.

Berenbaum, M. 1985. Brementown revisited: interactions among allelochemicals in plants. pp. 139–169. *In*: G.A. Cooper-Driver, T. Swain and E.E. Conn. (eds). Chemically Mediated Interactions between Plants and Other Organisms. Recent Advances in Phytochemistry, vol 19. Springer, Boston, MA.

Canelas, T., E. Thomsen, D. McDermott, E. Sternberg, M.B. Thomas and E. Worrall. 2021. Spatial targeting of Screening + Eave tubes (SET), a house-based malaria control intervention, in Côte d'Ivoire: A geostatistical modelling study. PLOS Glob Public Health. 1(11): e0000030.

Chang, K.S., J.H. Tak, J.H., S.I. Kim, W.J. Lee and Y.J. Ahn. 2006. Repellency of Cinnamomum cassia bark compounds and cream containing cassia oil to *Aedes aegypti* (Diptera: Culicidae) under laboratory and indoor conditions. Pest Manage. Sci. 62: 1032–1038.

Chaubey, M.K. 2007. Insecticidal activity of *Trachyspermum ammi* (Umbelliferae), *Anethum graveolens* (Umbelliferae) and *Nigella sativa* (Ranunculaceae) essential oils against stored-product beetle *Tribolium castaneum* Herbst (Coleoptera: Tenebrionidae). Afr. J. Agric. Res. 2(11): 596–600.

Choochote, W., U. Chaithong, K. Kamsuk, A. Jitpakdi, P. Tippawangkosol, B. Tuetun, et al. 2007. Repellent activity of selected essential oils against Aedes aegypti. Fitoterapia. 78: 359–364.

Curtis, C., J. Lines, B. Lu and A. Renz. 1989. Chapter 4—Natural and synthetic repellents. pp. 75–92. *In:* Curtis, C.F. (ed.). Appropriate Technology in Vector Control. CRC Press, Florida.

Deka, B., A. Babu, C. Baruah and M. Barthakur. 2021. Nanopesticides: A systematic review of their prospects with special reference to tea pest management. Front. Nutr. 8: 686131.

Dethier, V.G. 1956. Repellents. Annu. Rev. Entomol. 1: 181–202.

Elsayed, G.A. and A.G. Hassabo. 2021. Insect repellent of cellulosic fabrics (a review). Letters in Appl. Nano. Bio. Sci. 11.

EPA, 1997. Climate Change and Public Health.

Erler, F., I. Ulug and B. Yalcinkaya. 2006. Repellent activity of five essential oils against *Culex pipiens*. Fitoterapia. 77: 491–494.

FAO. 2018. Flavours and Fragrances of Plant Origin. Rome.

Foster, B.C., J.T. Arnason and C.J. Briggs. 2005. Natural health products and drug disposition. Annu. Rev. Pharmacol. Toxicol. 45: 203–226.

Fradin, M.S. and J.F. Day. 2002. Comparative efficacy of insect repellents against mosquito bites. N. Engl. J. Med. 347: 13–18.

García, M., O.J. Donadel, C.E. Ardanaz, C.E. Tonn and M.E. Sosa. 2005. Toxic and repellent effects of *Baccharis salicifolia* essential oil on *Tribolium castaneum*. Pest Manag. Sci. 61(6): 612-618. doi: 10.1002/ps.1028. PMID: 15714462.

Gbolade, A.A., A.O. Oyedele, M.B. Sosan, F.B. Adewoyin and O.L. Soyelu. 2000. Mosquito repellent activities of essential oils from two Nigerian *Ocimum* species. J. Trop. Med. Plants 1: 146–148.

Gillij, Y.G., R.M. Gleiser and J.A. Zygadlo. 2008. Mosquito repellent activity of essential oils of aromatic plants growing in Argentina. Bioresour. Technol. 99: 2507–2515.

Goel, D., Goel, R., Singh, V., Ali, M., Mallavarapu, G., Kumar, S., 2007. Composition of the essential oil from the root of Artemisia annua. J. Nat. Med. 61: 458–461.

Govere, J., D.N. Durrheim, N. Du. Toit, R.H. Hunt and M. Coetzee. 2000. Local plants as repellents against *Anopheles arabiensis*, in Mpumalanga province, South Africa. Cent. Afr. J. Med. 46(8): 213–216.

Guenther, E., 1972. The Essential Oils. Krieger Publishing Company, Florida, USA.

Hadis, M., M. Lulu, Y. Mekonnen and T. Asfaw. 2003. Field trials on the repellent activity of four plant products against mainly Mansonia population in Western Ethiopia. Phytother. Res. 17: 202–205.

Hori, M. 2003. Repellency of essential oils against the cigarette beetle, *Lasioderma serricorne* (Fabricius) (Coleoptera: Anobiidae). Appl. Entomol. Zool. 38: 467–473.

Hummelbrunner, L.A. and M.B. Isman. 2001. Acute, sublethal, antifeedant and synergistic effects of monoterpenoid essential oil compounds on the tobacco cutworm, *Spodoptera litura* (Lep. Noctuidae). J. Agric. Food Chem. 49: 715–720.

Isman, M.B. 2006. Botanical insecticides, deterrents, and repellents in modernagriculture and an increasingly regulated world. Annu. Rev. Entomol. 51: 45–66.

Jacobson, M. 1966. Chemical insects attractants and repellents. Annu. Rev. Entomol. 11: 403-422.

Jaenson, T.G.T., K. Palsson and A.K. Borg-Karlson. 2006. Evaluation of extracts and oils of mosquito (Diptera: Culicidae) repellent plants from Sweden and Guinea-Bissau. J. Med. Entomol. 43: 113–119.

Junkum, A., W. Maleewong, A. Saeung, D. Champakaew, A. Chansang, D. Amornlerdpison and B. Pitasawat. 2021. *Ligusticum sinense* nanoemulsion gel as potential repellent against *Aedes aegypti*, *Anopheles minimus*, and *Culex quinquefasciatus* (Diptera: Culicidae). Insects, 12: 596.

Kamsuk, K., W. Choochote, U. Chaithong, A. Jitpakdi, P. Tippawangkosol, D. Riyong, et al. 2006. Effectiveness of *Zanthoxylum piperitum*-derived essential oil as an alternative repellent under laboratory and field applications. Parasitol. Res. 100: 339–345.

Katerinopoulos, H.E., G. Pagona, A. Afratis, N. Stratigakis and N. Roditakis. 2005. Composition and insect attracting activity of the essential oil of *Rosmarinus officinalis*. J. Chem. Ecol. 31: 111–122.

Katz, T., J. Miller and A. Hebert. 2008. Insect repellents: historical perspectives and new developments. J. Am. Acad. Dermatol. 58: 865–871.

Khan, A.A., H.I. Maibach and D.L. Skidmore. 1975. Addition of vanillin to mosquito repellents to increase protection time. Mosq. News. 35: 223–225.

Kiran, S. and P. Devi. 2007. Evaluation of mosquitocidal activity of essential oil and sesquiterpenes from leaves of *Chloroxylon swietenia* DC. Parasitol. Res. 101: 413–418.

Kline, D.L., K. McKenzie and A. Bowman. 2022. Semi-field evaluation of arthropod repellents: emphasis on spatial repellents. Adv. Arthropod. Repel. 193–236.

Koschier, E. and K. Sedy. 2003. Labiate essential oils affecting host selection and acceptance of *Thrips tabaci* lindeman. Crop. Prot. 22: 929–934.

Landolt, P.J., R.W. Hofstetter and L.L. Biddick. 1999. Plant essential oils as arrestants and repellents for neonate larvae of the codling moth (Lepidoptera: Tortricidae). Environ. Entomol. 28: 954–960.

Li, M.X., Y.P. Ma, H.X. Zhang, H.Z. Sun, H.H. Su, S.J. Pei, et al. 2021. Repellent, larvicidal and adulticidal activities of essential oil from Dai medicinal plant Zingiber cassumunar against Aedes albopictus. Plant Divers, 43: 317–323.

Liu, C.H., A.K. Mishra, R.X. Tan, C. Tang, H. Yang and Y.F. Shen. 2006. Repellent and insecticidal activities of essential oils from *Artemisia princeps* and *Cinnamomum camphora* and their effect on seed germination of wheat and broad bean. Bioresour. Technol. 97: 1969–1973.

Maji, T., I. Baruah, S. Dube and M.R. Hussain. 2007. Microencapsulation of *Zanthoxylum limonella* oil (ZLO) in glutaraldehyde crosslinked gelatin for mosquito repellent application. Bioresour. Technol. 98: 840–844.

Moore, S.J., N. Hill, C. Ruiz and M.M. Cameron. 2007. Field evaluation of traditionally used plant-based insect repellents and fumigants against the malaria vector Anopheles darlingi in Riberalta, Bolivian Amazon. J. Med. Entomol. 44: 624–630.

Ndungu, M.W., S.C. Chhabraa and W. Lwandeb. 1999. Cleome hirta essential oil as livestock tick (*Rhipicephalus appendiculatus*) and maize weevil (*Sitophilus zeamais*) repellent. Fitoterapia. 70: 514–516.

Odalo, J.O., M.O. Omolo, H. Malebo, J. Angira, P.M. Njeru, I.O. Ndiege, et al. 2005. Repellency of essential oils of some plants from the Kenyan coast against Anopheles gambiae. Acta. Trop. 95: 210–218.

Omolo, M.O., D. Okinyo, I.O. Ndiege, W. Lwande and A. Hassanali. 2004. Repellency of essential oils of some Kenyan plants against Anopheles gambiae. Phytochemistry. 65: 2797–2802.

Oyedele, D.J., P. Schjønning, E. Sibbesen and K. Debosz. 1999. Aggregation and organic matter fractions of three Nigerian soils as affected by soil disturbance and incorporation of plant material. Soil Tillage Res. 50: 105–114.

Oyedele, A.O., A.A. Gbolade, M.B. Sosan, F.B. Adewoyin, O.L. Soyelu and O.O. Orafidiya. 2002. Formulation of an effective mosquito-repellent topical product from Lemongrass oil. Phytomedicine 9: 259–262.

Pereira, F., A.A. Pessoa, G.C.Á. Yamaguchi, L.F. Stanton, M.A. Serravite, A.M. Pereira, et al. 2021. Larvicidal activity of essential oils from piper species against strains of *Aedes aegypti* (Diptera: Culicidae) resistant to pyrethroids. Front. Plant. Sci. 12.

Pitasawat, B., W. Choochote, B. Tuetun, P. Tippawangkosol, D. Kanjanapothi, A. Jitpakdi, et al. 2003. Repellency of aromatic turmeric Curcuma aromatica under laboratory and field conditions. J. Vector Ecol. 28: 234–240.

Prajapati, V., A.K. Tripathi, K.K. Aggarwal and S.P.S. Khanuja. 2005. Insecticidal, repellent and oviposition-deterrent activity of selected essential oils against *Anopheles stephensi*, *Aedes aegypti* and *Culex quinquefasciatus*. Bioresour. Technol. 96: 1749–1757.

Pushpanathan, T., A. Jebanesan and M. Govindarajan. 2006. Larvicidal, ovicidal and repellent activities of *Cymbopogan citratus* Stapf. (Graminae) essential oil against the filarial mosquito *Culex quinquefasciatus* (Say) (Diptera: Culicidae). Trop. Biomed. 23: 208–212.

Schreck, C.E. 1977. Techniques for the evaluation of insect repellents: a critical review. Annu. Rev. Entomol. 22: 101–119.

Sukumar, K., M.J. Perich and L.R. Boobar. 1991. Botanical derivatives in mosquito control: a review. J. Am. Mosq. Control Assoc. 7: 210–237.

Tawatsin, A., S.D. Wratten, R.R. Scott, U. Thavara and Y. Techandamrongsin. 2001. Repellency of volatile oils from plants against three mosquito vectors. J. Vector Ecol. 26: 76–82.

Toloza, A.C., J. Zygadlo, C.G. Mougabure, F. Biurrun, E. Zerba and M.I. Picollo. 2006. Fumigant and repellent properties of essential oils and component compounds against permethrin-resistant *Pediculus humanus capitis* (Anoplura: Pediculidae) from Argentina. J. Med. Entomol. 43: 889–895.

Toloza, A.C., A. Lucia, E. Zerba, H. Masuh and M.I. Picollo. 2008. Interspecific hybridization of eucalyptus as a potential tool to improve the bioactivity of essential oils against permethrin-resistant head lice from Argentina. Bioresour. Technol. 99: 7341–7347.

Trongtokit, Y., Y. Rongsriyam, N. Komalamisra and C. Apiwathnasorn. 2005. Comparative repellency of 38 essential oils against mosquito bites. Phytother. Res. 19: 303–309.

Trumble, J.T. 2002. Caveat emptor: safety considerations for natural products used in arthropod control. Am. Entomol. 48: 7–13.

Tunón, H., W. Thorsell, A. Mikiver and I. Malander. 2006. Arthropod repellency, especially tick (Ixodes ricinus), exerted by extract from *Artemisia abrotanum* and essential oil from flowers of *Dianthus caryophyllum*. Fitoterapia. 77: 257–261.

Wang, J., F. Zhua, X.M. Zhoua, C.Y. Niua and C.L. Lei. 2006. Repellent and fumigant activity of essential oil from *Artemisia vulgaris* to *Tribolium castaneum* (Herbst) (Coleoptera: Tenebrionidae). J. Stored Prod. Res. 42: 339–347.

Wang, Z., J. Song, J. Chen, Z. Song, S. Shang, Z. Jiang, et al. 2008. QSAR study of mosquito repellents from terpenoid with a six-member-ring. Bioorg. Med. Chem. Lett. 18: 2854–2859.

Yang, Y.C., E.H. Lee, H.S. Lee, D.K. Lee and Y.J. Ahn. 2004. Repellency of aromatic medicinal plant extracts and a steam distillate to Aedes aegypti. J. Am. Mosq. Control Assoc. 20: 146–149.

Zygadlo, J.A. and H.R. Juliani. 2003. Study of essential oil composition of aromatic plants from Argentina. pp. 273–293. In: D.K. Majundar, J.N. Govil and V.K. Singh (eds). Recent Progress in Medicinal Plants, vol. 8. Studium Press, Houston, TX.

7

Essential Oils in Plant Defense Priming

María Fe Andres[1]*, Marta Berrocal-Lobo*[2]*, Sabrina Kesraoui*[1]*,*
Serine Soudani[2] *and Azucena González-Coloma*[1*]

Introduction to Essential Oils

Essential Oils (EOs) are volatile extracts obtained mainly from aromatic and medicinal plants. The *Lamiaceae* is the most important EO producing plant family and includes about 6900–7200 species belonging to 236 genera distributed all over the world, with the Mediterranean and temperate regions predominating (Tamokou et al. 2017). EOs are an important source of biologically active compounds that have antibacterial, insecticidal, fungicidal, nematicidal, herbicidal, antioxidant, and anti-inflammatory activities (Bassole and Juliani 2012, Turek and Stintzing 2013).

Essential oils are mostly composed of volatile terpenes (Bassole and Juliani 2012), are usually comprised of 20 to 60 substances, and characterized in many cases by up to three major components in a relatively high concentration, while other compounds are present in trace amounts (Bassole and Juliani 2012, Bakkali et al. 2008). EO components can be grouped into two main groups (Figure 7.1):

- Terpene hydrocarbons, constituting monoterpenes (representing 80% of the EO's composition) and sesquiterpenes (Turek and Stintzling 2012, Asbahani et al. 2015).
- Oxygenated compounds, composed mainly of alcohols, phenols, aldehydes, and esters. The aromatic and oxygenated compounds are less abundant than terpenes in EO (Bakkali et al. 2008, Buckle 2014).

Human beings have long used Medicinal and Aromatic Plants (MAPs) to produce EOs, mainly as flavours and fragrances based on the traditional use of aromatic plants as culinary herbs and spices. More recently, their use has expanded to include use in human medicine, as phytopharmaceuticals, and in aromatherapy (Isman 2020a).

Many of these volatile substances have not only physiological functions, but also play multiple ecological functions. They can act as internal messengers, as

[1] Instituto de Ciencias Agrarias, Consejo Superior de Investigaciones Científicas (CSIC), Madrid, Spain.
School of Forestry Engineering and Natural Resources, Polytechnical University of Madrid, Spain.

* Corresponding author: azu@ica.csic.es

defences against herbivores, or as volatiles that also attract pollinating insects to their host (Hedden et al. 2002).

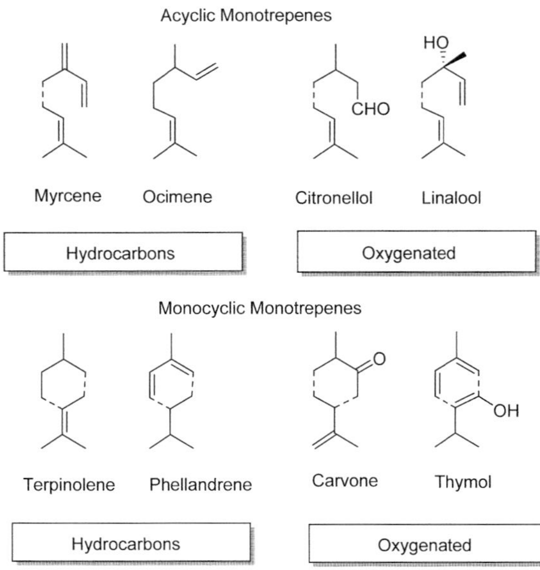

Figure 7.1 The main components responsible for the biological properties of Essential Oils.

Lately, interest in essential oils has focused on their bioactivity acting as biocontrol agents, used at low concentrations, against pests and disease-causing organisms, and thus their potential use as alternatives to synthetic chemical pesticides as a means for crop protection and pest control. This expansion of the potential uses of EOs has intensified academic and industrial research on the biological activities of these plant compounds (Isman 2020b).

Biocontrol is a widely used concept that has recently gained interest in the impartial control of plant pests (Ravensberg 2015). With the increasing recognition of Integrated Pest-Management approaches (IPM) (Barzman et al. 2015), the obtaining of environmental-friendly biopesticides is an excellent alternative to synthetic and contaminant chemicals and molecules (Prabha et al. 2016), due to their toxicity and poor biodegradability, and to achieve more sustainable production methods that meet the consumers and societal expectations (Werrie et al. 2020). The EU framework aims to achieve sustainable use of plant protection products by promoting integrated pest management (*Sustainable use of plant protection products—European court of auditors, 2020*). In this context, numerous results have described the strong biopesticidal potential of EOs due to their antifungal (Arraiza et al. 2018), insecticidal (Isman 2020a), or nematicidal activities (Andrés et al. 2012, Werrie et al. 2020).

However, their direct use as biopesticides has shown problems such as phytotoxicity, which has long been considered a major obstacle to their development into biopesticides (insecticides, fungicides, etc.), their potential impact on the organoleptic quality of the resulting food, and the quantities needed (Isman 2020b). Additionally, EOs suffer a loss of efficiency when used directly in the field, mainly due to their volatile nature and susceptibility to degradation (Raveau et al. 2020). These natural properties of EOs make necessary research work to improve the knowledge has about EOs and how to apply them for biocontrol of plant disease.

In this chapter, the focus is on current knowledge which one has about EOs and what are the lastest approaches that are being applied in this growing area of research.

Cultivation and Domestication of MAPs to Produce EOs

The chemical composition of essential oils varies significantly from one region to another (Aghmouchi et al. 2018). EOs from MAPs are usually extracted from plants collected during the flowering period, before the seeds germinate, which could cause a reduction in the regeneration of these plants. This fact makes it difficult to get high efficiency for obtaining them and highlights the importance of the domestication of MAPs (Abbad et al. 2011, Kesraoui et al. 2022).

MAPs are sometimes collected without proper control resulting in the loss of habitat. This puts pressure on wild populations of medicinal plants whose disappearance has accelerated, especially in developing countries such as India, China, Nepal, Kenya, Tanzania, and Uganda (Chen et al. 2016). Therefore, with the increasing demand for standardized, homogeneous raw material in industrial societies, numerous wild MAPs species have been domesticated and systematically cultivated.

Domestication provides several advantages over wild harvest for the production of essential oils, it helps to avoid the admixture and adulteration through reliable botanical identification, gives better control over harvested quantities, and facilitates the selection of genotypes with desirable characteristics, especially quality. Additionally, domestication controls the influence on the history of the plant material and post-harvest handling (Schippmann et al. 2006, Franz et al. 2011, Navarro-Rocha et al. 2020). The key initial steps are MAP, the exact botanical identification, and a detailed description of the growing area. At the same time, it is necessary to carry out a phytochemical evaluation of the initially collected plant material to identify the chemotypes (Franz and Novak 2015). The domestication process could affect the chemical composition of MAPs and consequently affect their biological activities. For this reason, understanding the influence of domestication is critical to conduct the cultivation of MAP species (Yousif et al. 2021). Numerous examples of successful MAPs domestication include *Origanum* sp. (Kitiki 1997, Putievsky et al. 1997, Ceylan et al. 1999, Yousif et al. 2021), *Lippia sp.* (Fischer et al. 2004*), Hyptis suaveolens* (Grassi 2003), and *T. lucida* (Goehler 2006), *Artemisia absinthium* (Julio et al. 2015) *Lavandula luisieri* (Julio et al. 2016), *Mentha sp.* (Vining et al. 2020) and *Satureja montana* (Navarro-Rocha et al. 2020).

Additional problems to be considered in MAP crop production are contamination with heavy metals, damage by pests and diseases, and pesticide residues. Therefore, quality assurance measures are needed to ensure that plants are produced with care so that the negative impacts during wild collection, cultivation, processing, and storage can be limited. Guidelines for good agricultural practices and standards for Sustainable Wild Collection have been established (World Health Organization 2003, Schippmann et al. 2006).

EOs in Plant Protection

Direct Effects

Antifungal Activity

Phytopathogenic fungi are responsible for nearly 30% of all crop diseases and could have a big impact on crops, affecting them during cultivation or post-harvest, during storage (Raveau et al. 2020). Past research has suggested evidence of how EOs act in the fungal cells (Duduk et al. 2015, Hong et al. 2015, Karimi et al. 2016, Parra Amin et al. 2021, Kalhoro et al. 2021, Li et al. 2021), via assays assessed from cell wall alterations (de Billerbeck et al. 2001, Scorzoni et al. 2017), or gene expression modifications diminishing fungal virulence (Reuveni et al. 2020, Li et al. 2021, Mani-López et al. 2021).

The effects of EOs against a wide range of fungal species have been extensively studied *in vitro* but not their mechanisms of action. A recent review (Raveau et al. 2020) about the potential of EOs for the biocontrol of phytopathogens pointed out different mechanisms regarding antifungal properties (Raveau et al. 2020). The inhibition of the fungi cell-wall formation, by a *Cinnamomum zeylanicum* essential oil has been reported to exert antifungal activity in *Candida albicans.* This essential oil triggered the cell cycle arrest by disrupting beta-tubulin distribution, leading to mitotic spindle defects, ultimately compromising the cell membrane, and allowing leakage of cellular components (Shahina et al. 2018). Citral, an EO component, was tested on *Penicillium italicum*, the results obtained from RNA-seq showed its effect on the inhibition of ergosterol biosynthesis by affecting the expression of *ERG6* gene responsible for the synthesis of ergosterol in pathogens (Jensen-Pergakes et al. 1998, Ou Yang et al. 2016). The decrease in ergosterol synthesis affects the fungal mitochondria by inhibiting the mitochondrial electron transport. An onion EO, caused depolarization of mitochondrial membranes by lowering the membrane potential, affecting the ionic Ca++ circuit and ion channels, and the proton pump and ATP pool, decreasing the pH gradient (Wu et al. 2006, Bakkali et al. 2008). A decrease in energy metabolism could lead to a slowdown in transcription and translation, consistent with slower ribosome biogenesis and faster RNA degradation (Li et al. 2021). *Melaleuca cajuputi* EO reduced the MIC value (the lowest concentration of drug showing no visible growth) of fluconazole and the expression of *MDR1*, a gene encoding drug-efflux pumps in *Candida albicans* (Keereedach et al. 2020).

Nematicidal Activity

Plant-parasitic nematodes are the most destructive group of plant pathogens worldwide and their control is extremely challenging (Bird et al. 2009). Therefore, in the last decade, a great deal of effort has been focused on the study of the nematicidal activity of EOs and their constituents as potential sources of commercial products for the management of the root-knot nematodes, *Meloidogyne* spp, one of the most economically damaging genera on horticultural and field crops.

Many EOs extracted from different botanical families has been analyzed *in vitro* for nematicidal activity mainly against *Meloidogyne spp.* (Andres et al. 2012). Among EO-producing plants, some families such as Lamiaceae, Asteraceae, Myrtaceae, Rutaceae, Lauraceae, and Poaceae have been widely studied. The EOs from MAPs, especially of the genera *Artemisia, Cympogon, Lavandula, Mentha, Origanum, Ocimum, Satureja, Thymus,* and aromatic trees of the genera *Citrus, Eucalyptus,* and *Eugenia* whose nematicidal effects the root-knot nematodes (*M. Arenariaa, M. chitwoodi, M. hapla, M. incognita,* and *M. javanica*), have been reported in depth (Andres et al. 2012, Faria et al. 2016, Jeon et al. 2016, Patidar et al. 2016, Ozdemir and Gozel 2017, Ozdemir and Gozel 2018, Barros et al. 2019, Eloh et al. 2020, Felek et al. 2019).

The mode of action of EOs and their constituents is of practical importance for nematode control, as it could give useful information on the most appropriate formulation and the means of delivery. Neurotoxic effects on nematodes of EO components have been reported, involving several mechanisms, particularly through GABA, octopamine synapses, and acetylcholinesterase inhibition (Andres et al. 2012). Recently, active EOs has been investigated to correlate their mechanism of action for target-specific binding affinities toward nematode proteins. Molecular modelling and in silico studies suggest a higher binding capacity of geraniol, b-terpineol, citronellal, l-limonene, g-terpinene, α-bulnesene, and α-guaiene to the selected target proteins (ODR1 and ODR3 odorant response genes) and AChE, (Keerthiraj et al. 2021, Kundu et al. 2021). The insight on biochemical ligand-target protein interactions could be helpful in the selection of biomolecules and essential oils for the development of practically viable bionematicidal products (Keerthiraj et al. 2021).

Insecticidal Activity

The insecticidal action of plant essential oils (EOs) has been an area of intensive research lately. According to recent bibliometric analyses, more papers have been published in recent years on this group of natural insecticidal materials than on any other type of chemical class of plant-derived natural products (Isman 2020b, Isman and Grieneisen 2014).

The toxic and sublethal behavioural effects observed in insects and related arthropods can be attributed to mono- and sesquiterpenoids present in essential oils (Isman and Tak 2017, Isman 2020a). Low molecular weight terpenoids can inhibit acetylcholine esterase (AChE) enzyme activity in laboratory bioassays, but such bioactivity seldom appears to correlate with toxicity *in vivo* in target insect species (Isman and Tak 2017). Recently, a decrease in AChE activity was observed following the exposure of *Alphitobius diapering* to EO of *Illicium verum* and correlated with the loss of the refuge-seeking capacity and the loss of locomotor capacity (Peter et al. 2022). The inhibition of the catalytic activity of acetylcholinesterase (AChE), Glutathione S-Transferase (GST), and catalase (CAT) in Tribolium *castaneum* has been reported for EOs from *Piper nigrum and Rosmarinus officinalis. Citrus sinensis, Piper aduncum,* and *Zanthoxylum monophyllum* inhibited GST activity and *Lavandula angustifolia, C. sempervirens,* and *Eucalyptus* sp inhibited CAT activity (Oviedo-Sarmiento et al. 2021).

Receptors for octopamine, a neuromodulator in arthropods, have been identified as a target for some of these terpenoids (Enan 2001, Kostyukovsky et al. 2002, Price and Berry 2006). Other targets in the insect-nervous system include GABA-gated chloride channels (Priestley et al. 2003, Tong and Coats 2010) and the nicotinic acetylcholine receptor (Tong et al. 2013).

Plant EOs are often complex mixtures of terpenoids and their bioactivity can be the result of synergy among constituents (Tak and Isman 2015, 2017). Synergy among the major constituents of rosemary (*Rosemarinus officinalis*) and lemongrass (*Cymbopogon citratus*) EOs, resulted from increased penetration of toxicants through the insect's integument (Tak and Isman 2016, Tak et al. 2017). A fumigant experiment against *Triboleum castaneum* with a total of 23 EOs showed the highest fumigant potential for EOs with greater diversity in their composition (Kim et al. 2021). Furthermore, mixtures in composition can reduce the development of resistance (Feng and Isman 1995) and behavioural habituation to deterrents (Akhtar and Isman 2003, Akhtar et al. 2012).

Toxic effects of EOs can upregulate physiologically important proteins and enzymes in insects. *Mentha arvensis* EO caused high contact toxicity in *Sitophilus granarius* adults and induced dramatic physiological changes in exposed insects revealed by the quantitative proteomics analysis. Most of the Differentially Expressed Proteins (DEPs) were upregulated and related to the development and functioning of the muscular and nervous systems, cellular respiration, protein synthesis, and detoxification (Renoz et al. 2022). In insecticide-resistant insects, EOs can synergize insecticide toxicity by inhibiting detoxification enzymes. Topical bioassays with binary mixtures of deltamethrin and individual EOs or their major constituents on delthametrin resistant and susceptible bed bugs (*Cimex lectularius*) caused significant increases in mortality through inhibition of P450 activity by EO constituents in resistant bed bugs (Gaire et al. 2020, 2021).

Indirect Effects—Priming

Several articles have been published characterizing molecular plant response to inorganic and organic chemicals, plant elicitors, and Pathogen or Disease-Associated Molecular Patterns (PAMPs and DAMPs, respectively). However, a smaller number of studies have been conducted to study the contribution of essential oils to combat phytopathogens at the transcriptomic and or metabolomic level (Arraiza et al. 2018). EOs can act as priming molecules both in biotic and abiotic plant-stress responses (Bertrand et al. 2021), and is an effective and sustainable tool to control seed-borne diseases (Klein et al. 2017, Spadaro et al. 2017). The metabolomic approach is being used recently for the characterization of metabolic pathways affected by plant priming (Bertrand et al. 2021).

The Table 7.1 shows the most recent priming works using EOs to induce biotic stress tolerance conducted since 2013, sequentially and accordingly updated (Mustafa et al. 2019).

Table 7.1 List of essential oils with reported primming effects

Disease	Plant essential oil	Primed species	References
Colletotrichum gloeosporioides	*Thymus vulgaris*	Avocado	Sellamuthu et al. 2013
Botrytis cinerea Fusarium wilt	*Thymus capitatus*	Tomato	Ben-jabeur et al. 2015
Colletotrichum gloeosporioides	Ginger and cinnamon	Mango	Gerefa et al. 2015
Colletotrichum gloeosporioides	Cinnamon oil, clove oil	Pepper fruit	Hong et al. 2015
Colletotrichum acutatum	Citronella	Chilli	Dela Cueva and Balendres. 2018
Colletotrichum gloeosporioides	Savory	Papaya	Sarkhosh et al. 2018
Botrytis cinerea	Thyme and savory	Apple	Banani et al. 2018
Phakopsora pachyrhizi	Candy mint peppermint	Soybean (*Glycine max*)	Sukegawa et al. 2018
Colletotrichum musae	*Thymus vulgaris*	Bananas	Vilaplana et al. 2018
Plasmopara viticola	Oregano	Grapevine (*Vitis Vinifera*)	Rienth et al. 2019
Colletotrichum nymphaeae	Thyme, cinnamon bark, and clove	Strawberry	Duduk et al. 2015
Colletotrichum nymphaeae *Fusarium solani*	*Anethum graveolens*	Strawberry Bean (*Phaseolus vulgaris L.*)	Karimi et al. 2016
	Allium sativum, Rosmarinus officinali		Hosseini et al. 2020
	Lemongrass		Eke et al. 2020
Mycosphaerella fijiensis	Tea tree (*Melaleuca alternifolia*)	Bananas	Reuveni et al. 2020
Colletotrichum acutatum	Cinnamon, basil, and lemongrass	Mango	Danh et al. 2021

The avocado fruit exposed to thyme essential oil had higher antioxidant enzyme activities (SOD, POD, and CAT) than the untreated control fruit (Sellamuthu et al. 2013). Thyme essential oil was evaluated in its capacity to protect tomato seedlings by the accumulation of peroxidases, which are the first line of defence against ROS (Ben-jabeur et al. 2015). Based on gene expression results, it is postulated that one mechanism by which thyme effectively controls grey mould disease in apple fruit is by inducing the PR-8 gene (Banani et al. 2018).

Mint volatiles triggered conserved signalling pathways in soybean plants and promoted histone modifications of defence genes, including TI and PR1 (Sukegawa et al. 2018). The vapour of *Oreganum vulgare* EO triggered a multi-layered immune system in grapevine. The analysis of gene expression revealed a complex activation of hormonal interactions involving JA, ET and SA biosynthesis and their signalling cascades (Rienth et al. 2019). *Allium sativum* and *Rosmarinus officinali* EOs preserved the quality parameters in treated strawberry fruits against *Colletotrichum nymphaeae* due to an increase in their phenolic content and the activity of the defence related enzymes such as peroxidase (Hosseini et al. 2020).

Recently, direct, and indirect plant protection effects of *Artemisia absinthium* essential oil on tomato seedlings against *Fusarium oxysporum* have been demonstrated. The EO exhibited an *in vitro* antifungal effect. In addition, tomato seedlings germinated from seeds pretreated with the EO were protected against the fungus. The EO treatment increased callose deposition and the production of Reactive Oxygen Species (ROS) on seed surfaces and primed a durable defence. The metabolomic analysis of the EO treated seedlings showed the induction of vanillic acid, coumarin, lycopene, and oleamide in the presence of the pathogen. RNA-seq analysis suggests that AEO treatment could induce de novo epigenetic changes in the tomato, modulating the speed and extent of its immune response to *F. oxyxporum*. The EO-seed coating could be a new strategy to prime durable tomato resistance, compatible with other environmentally friendly biopesticides (Soudani et al. 2022).

Conclusion and Perspectives

Essential oils are active against plant pests and diseases (insects, fungi, nematodes…), have synergistic effects that lower the risk of resistance development, and therefore are considered ingredients for the development of new biopesticides. However, problems with biomass availability, chemical stability, formulation, and phytotoxicity among others are the major obstacles to their development into biopesticides. Plant domestication and cultivation can partially solve the problems related to chemical stability (chemotypes) and biomass availability and new formulations can be used to overcome volatility and stability problems.

Recently, the priming effects of EOs on plants have been demonstrated opening new application opportunities through indirect effects. Most priming studies have been carried out with EO treated plants and fungal pathogens. Recently, a new application method (seed coating with EO) has shown priming effects in tomato seedlings against a fungal pathogen (*Fusarium oxysporum*) involving metabolic and epigenetic changes in the plant. The main advantage of this application is the low amounts required for a long-term effect and opens a new field for the development of EO-based biopesticides. However, more research is needed to determine the specificity of the plant response to the EO composition using these new strategies to improve plant tolerance to pest species.

Acknowledgements

This work was supported by grants ERASMUS+ European Hub on New Challenges in the Field of Essential Oils (EOHUB) and PID2019-106222RB-C31/SRA (State Research Agency, 10.13039/501100011033). SK and SS were funded by an international PhD fellow from the UPM-Algerian Government.

References

Abbad, A., R. Belaqziz, K. Bekkouche and M. Markouk. 2011. Influence of temperature and water potential on laboratory germination of two Moroccan endemic thymes: *Thymus maroccanus* Ball. and *Thymus broussonetii* Boiss. Afr. J. Agri. Res. 6: 4740–4745.

Aghmouchi, Y., O. Belmehdi, N.S. Snehaji and J. Abrini. 2018. Chemical composition and antibacterial activity of *Origanum compactum* Benth. essential oils from different areas at northern Morocco. S. Afri. J. Bot. 115: 120–125.

Akhtar, Y. and M.B. Isman. 2003. Binary mixtures of feeding deterrents mitigate the decrease in feeding deterrent response to antifeedants following prolonged exposure in the cabbage looper, *Trichoplusia ni* (*Lepidoptera: Noctuidae*). Chemoecol 13: 177–182.

Akhtar, Y., E. Pages, A. Stevens, R. Bradbury, C.A.G. da Camara and M.B. Isman. 2012. Effect of chemical complexity of essential oils on feeding deterrence in larvae of the cabbage looper. Physiol. Entomol. 37: 81–91.

Andrés, M.F., A. González-Coloma, J. Sanz, J. Burillo and P. Sainz. 2012. Nematicidal activity of essential oils: a review. Phytochem Rev. 11: 371–390.

Arraiza, M.P., A. González-Coloma, M.F. Andrés, M. Berrocal-Lobo, J.A. Domínguez-Núñez, A.C. Da Costa, et al. 2018. Antifungal effect of essential oils. *In*: H.E. El-Shemy (ed.). Potential of Essential Oils, IntechOpen. London. UK. doi: 10.5772/intechopen.78008

Asbahani, A.E., K. Miladi, W. Badri, M. Sala, E.H.A. Addi, H. Casabianca, et al. 2015. Essential oils: From extraction to encapsulation. Int. J. Pharm. 483: 220–243.

Bakkali, F., S. Averbeck, D. Averbeck and M. Idaomar. 2008. Biological effects of essential oils—A review. Food Chem. Toxicol. 46: 446–475.

Banani, H., L. Olivieri, K. Santoro, A. Garibaldi, M.L. Gullino and D. Spadaro. 2018. Thyme and savory essential oil efficacy and induction of resistance against *Botrytis cinerea* through priming of defense responses in apple. Foods. 7: 11.

Barros, A., V. Campos, L. de Paula, D. Oliveira, F. de Silva, W. Terra, et al. 2019. Nematicidal screening of essential oils and potent toxicity of *Dysphania ambrosioides* essential oil against *Meloidogyne incognita in vitro* and *in vivo*. J. Phytopathol. 167: 380–389.

Barzman, M., P. Bàrberi, A.N.E. Birch, P. Boonekamp, S. Dachbrodt-Saaydeh, B. Graf, et al. 2015. Eight principles of integrated pest management. Agron. Sustain. Dev. 35: 1199–1215.

Bassolé, I.H.N. and H.R. Juliani. 2012. Essential oils in combination and their antimicrobial properties. Molecules. 17: 3989–4006.

Ben-jabeur, M., G. Emna, M. Machraoui and W. Hamada. 2015. Thyme essential oil as a defense inducer of tomato against gray mold and *Fusarium* wilt. Plant Physiol. Biochem. 94: 35–40.

Bird, D.Mc.K, V.M. Williamson, P. Abad, J. McCarter and E.G.J. Danchin. 2009. The genome of root-knot nematodes. Annu. Rev. Phytopathol. 47: 333–351.

Buckle, J. 2014. Essential oils in practice. Clinical aromatherapy-e-book. Churchill Livingstone, Edinburgh, Scotland. UK.

Bertrand, C., A. Gonzalez-Coloma and C. Prigent-Combaret. 2021. Plant metabolomics to the benefit of crop crotection and growth stimulation. Adv. Bot. Res. 98: 107–132.

Ceylan, A., E. Bayram and H. Geren. 1999. Investigation on agronomic and quality characteristics of improved clones in *Origanum* (*Origanum onites* L.) breeding. Turk. J. Agric. For. 23: 1163–1168.

Chen, S.L., H. Yu, H.M. Luo, Q. Wu, C.F. Li and A. Steinmetz. 2016. Conservation and sustainable use of medicinal plants: problems, progress, and prospects. Chin. Med. 11: 37.

European Court of Auditors. 2020. Sustainable use of plant protection products: limited progress in measuring and reducing risks. Special report No. 05, 2020, Publications Office. ttps://data.europa.eu/doi/10.2865/349084.

Danh, L.T., B.T. Giao, C.T. Duong, N.T.T. Nga, D.T.K. Tien, N.T. Tuan, et al. 2021. Use of essential oils for the control of anthracnose disease caused by *Colletotrichum acutatum* on post-harvest mangoes of cat hoa loc variety. Membranes. 11: 719.

de Billerbeck, V.G., C.G. Roques, J.M. Bessière, J.L. Fonvieille and R. Dargent. 2001. Effects of *Cymbopogon nardus* (L.) W. Watson essential oil on the growth and morphogenesis of *Aspergillus niger*. Can. J. Microbiol. 47: 9–17.

Dela Cueva, F. and M.A. Balendres. 2018. Efficacy of citronella essential oil for the management of chilli anthracnose. Eur. J. Plant. Pathol. 152: 461–68.

Duduk, N., T. Markovic, M. Vasic, B. Duduk, I. Vico and A. Obradovic. 2015. Antifungal activity of three essential oils against *Colletotrichum acutatum*, the causal agent of strawberry anthracnose. J. Essent. Oil-Bear Plants. 18: 529–37.

Eke, P., S. Adamou, R. Fokom, V. Dinango Nya, P.V. Tsouh Fokou, L. Nana Wakam, et al. 2020. Arbuscular mycorrhizal fungi alter antifungal potential of lemongrass essential oil against *Fusarium solani*, causing root rot in common bean (*Phaseolus vulgaris* L.). Heliyon. 6: e05737.

Eloh, K., K. Kpegba, N. Sasanelli, H.K. Koumaglo and P. Caboni. 2020. Nematicidal activity of some essential plant oils from tropical West African. Int. J. Pest Manag. 66: 131–141.

Enan, E. 2001. Insecticidal activity of essential oils: octopaminergic sites of action. Comp. Biochem. Physiol. 130: 325–337.

Faria, J., I. Sena, B. Ribeiro, A. Rodrigues, C. Maleita, I. Abrantes, et al. 2016. First report on *Meloidogyne chitwoodi* hatching inhibition activity of essential oils and essential oils fractions. J. Pest Sci. 89: 207–217.

Felek, A.F., M.M. Ozcan and F. Akyazi. 2019. Effects of essential oils distilled from some medicinal and aromatic plants against root knot nematode (*Meloidogyne hapla*). J. Appl. Sci. Environ. Manage. 23: 1425–1430.

Feng, R. and M.B. Isman. 1995. Selectionf or resistance to azadirachtin in the green peach aphid, *Myzus persicae*. Experientia. 51: 831–833.

Fischer, U., R. Lopez, E. Pöll, S. Vetter, J. Novak, and C.M. Franz. 2004. Two chemotypes within *Lippia alba* populations in Guatemala. Flavour Frag. J. 19: 333-335.

Franz, C., R. Chizzola, J. Novak and S. Sponza. 2011. Botanical species being used for manufacturing plant food supplements (PFS) and related products in the EU member states and selected third countries. Food. Funct. 2: 720.

Franz, C. and J. Novak. 2015. Sources of Essential Oils from: Handbook of Essential Oils, Science, Technology, and Applications. CRC Press. Boca Raton. USA.

Gaire, S., C.D. Lewis, W. Booth, M.E. Scharf, W. Zheng, M.D. Ginzel, et al. 2020. Bed bugs, *Cimex lectularius* L., exhibiting metabolic and target site deltamethrin resistance are susceptible to plant essential oils. Pestic. Biochem. Physiol. 169: 104667.

Gaire, S., W. Zheng, M.E. Scharf and A.D. Gondhalekar. 2021. Plant essential oil constituents enhance deltamethrin toxicity in a resistant population of bed bugs (*Cimex lectularius* L.) by inhibiting cytochrome P450 enzymes. Pestic. Biochem. Physiol. 175: 104829.

Gerefa, S., N. Satheesh and G. Berecha. 2015. Effect of essential oils treatment on anthracnose (*Colletotrichum gloeosporioides*) disease development, quality and shelf life of mango fruits (*Mangifera indica* L.). American-Eurasian J. Agric. Environ. Sci. 15: 2160–2169.

Goehler, I., 2006. Domestikation von Medizinalpflanzen und Untersuchungen zur Inkulturnahme von *Tagetes lucida Cav*. Dissertation, an der Universität für Bodenkultur Wien, Wein, Austria.

Grassi, P., 2003. Botanical and chemical investigations in *Hyptis* spp. (*Lamiaceae*) in El Salvador. Dissertation, Universität Wien, Wein, Austria.

Hedden, P., P. Harrewijn, A.M. van Oosten and P.G.M. Piron. 2002. Natural terpenoids as messengers. A multidisciplinary study of their production, biological functions and practical applications. Ann. Bot. 20: 299–300.

Hong, J.K., H.J. Yang, H. Jung, D.J. Yoon, M.K. Sang and Y.C. Jeun. 2015. application of volatile antifungal plant essential oils for controlling pepper fruit anthracnose by *Colletotrichum gloeosporioides*. Plant Pathol. J. 31: 269–77.

Hosseini, S., J. Amini, M.K. Saba, K. Karimi and I. Pertot. 2020. Preharvest and postharvest application of garlic and rosemary essential oils for controlling anthracnose and quality assessment of strawberry fruit during cold storage. Front. Microbiol. 11: 1855.

Isman, M.B. and M.J. Grieneisen. 2014. Botanical insecticide research: many publications, limited useful data. Trends. Plant. Sci. 19: 140–145.

Isman, M.B. and J.H. Tak. 2017. Inhibition of acetylcholinesterase by essential oils and monoterpenoids: a relevant mode of action for insecticidal essential oils? Biopestic. Int. 13: 71–78.

Isman, M.B. 2020a. Commercial development of plant essential oils and their constituents as active ingredients in bioinsecticides. Phytochem. Rev. 19: 235–241.

Isman, M.B. 2020b. Bioinsecticides based on plant essential oils: a short overview. Z. Naturforsch C. J. Biosci. 75: 179–182.

Jensen-Pergakes, K.L., M.A. Kennedy, N.D. Lees, R. Barbuch, C. Koegel and M. Bard. 1998. Sequencing, disruption, and characterization of the *Candida albicans* sterol methyltransferase (ERG6) gene: drug susceptibility studies in erg6 mutants. Antimicrob Agents Chemother. 42: 1160–1167.

Jeon, J., H. Ko, S. Kim and J. Lee. 2016. Chemical compositions and nematicidal activities of essential oils on *Meloidogyne hapla* (Nematoda: Tylenchida) under laboratory conditions. Korean. J. Pestic. Sci. 20: 30–34.

Julio, L.F., J. Burillo, C. Giménez, R. Cabrera, C.E. Díaz, J. Sanz, et al. 2015. Chemical and biocidal characterization of two cultivated *Artemisia absinthium* populations with different domestication levels. Ind. Crops. Prod. 76: 787–92.

Julio, L.F., A.F. Barrero, M.M. Herrador del Pino, J.F. Arteaga, J. Burillo, M.F. Andres, et al. 2016. Phytotoxic and nematicidal components of *Lavandula luisieri*. J. Nat. Prod. 79: 261–6.

Kalhoro, M.T., H. Zhang, G.M. Kalhoro, Y. Faqir, F. Wang, F. Nabi, et al. 2021. Screening the antifungal properties of essential oils against *Phytophthora colocasiae*. Res Sq (preprint).

Karimi, K., M. Arzanlou and I. Pertot. 2016. Antifungal activity of the dill (*Anethum graveolens* L.) seed essential oil against strawberry anthracnose under *in vitro* and *in vivo* conditions. Arch. Phytopathol. 49: 554–66.

Keereedach, P., H. Karnjana and B. Khaemaporn. 2020. Antifungal activity of thai cajuput oil and its effect on efflux-pump gene expression in fluconazole-resistant *Candida albicans* clinical isolates. Int. J. Microbiol. 2020: 5989206.

Keerthiraj, M., A. Mandal, T.K. Dutta, S. Saha, A. Dutta, A. Singh, et al. 2021. Nematicidal and molecular docking investigation of essential oils from *Pogostemon cablin* ecotypes against *Meloidogyne incognita.* Chem. Biodiversity. 18: 2100320.

Kesraoui, S., M.F. Andrés, M. Berrocal-Lobo, S. Soudani and A. Gonzalez-Coloma. 2022. Direct and indirect effects of essential oils for sustainable crop protection. Plants. 11: 2144. https://doi.org/10.3390/plants11162144.

Kim, S., J. Yoon and J. Tak. 2021. Synergistic mechanism of insecticidal activity in basil and mandarin essential oils against the tobacco cutworm. J. Pest Sci. 94: 1119–1131.

Kitiki, A. 1997. Status of cultivation and use of oregano in Turkey. Proc. IPGRI Workshop Oregano, Italy 14: 122–132.

Klein, J.D., A. Firmansyah, N. Panga, W. Abu-Aklin, M. Dekalo-Keren, T. Gefe, et al. 2017. Seed treatments with essential oils protect radish seedlings against drought. AIMS Agric. Food. 2: 345–53.

Kostyukovsky, M., A. Rafaeli, C. Gileadi, N. Demchenko and E. Shaaya. 2002. Activation of octopaminergic receptors by essential oil constituents isolated from aromatic plants: possible mode of action against insect pests. Pest Managem. Sci. 58: 1101–1106.

Kundu, A., A. Dutta, A. Mandal, L. Negi, M. Malik, R. Puramchatwad, et al. 2021. A comprehensive *in vitro* and *in silico*: analysis of nematicidal action of essential oils. Front. Plant Sci. 11: 614143.

Li, X., M. Liu, T. Huang, K. Yang, S. Zhou, Y. Li, et al. 2021. Antifungal effect of nerol via transcriptome analysis and cell growth repression in sweet potato spoilage fungi *Ceratocystis fimbriata*. Postharvest Biol. Technol. 171: 111343.

Mani-López, E., O. Cortés-Zavaleta and A. López-Malo. 2021. A review of the methods used to determine the target site or the mechanism of action of essential oils and their components against fungi. SN Appl. Sci. 3: 1–25.

Mustafa, S., N. Ahmed, A. Saboor, S. Ahmad, S. Hussain, et al. 2019. Seed priming for disease resistance in plants. pp. 333–362. *In*: M. Hasanuzzaman and V. Fotopoulos (eds). Priming and Pretreatment of Seeds and Seedlings. Springer. Singapore.

Navarro-Rocha, J., M.F. Andres, C.E. Diaz, J. Burillo and A. Gonzalez-Coloma. 2020. Composition and biocidal properties of essential oil from pre-domesticated Spanish *Satureja montana*. Ind. Crop. Prod. 145: 111958.

Ou Yang, Q., N. Tao and G. Jing. 2016. Transcriptional profiling analysis of *Penicillium digitatum,* the causal agent of citrus green mold, unravels an inhibited ergosterol biosynthesis pathway in response to citral. BMC Genom. 17: 599.

Oviedo-Sarmiento, J.S, J.J. Bustos Cortes, W.A. Delgado Ávila, L.E. Cuca Suárez, E. Herrera Daza, O.J. Patiño-Ladino, et al. 2021. Fumigant toxicity and biochemical effects of selected essential oils toward the red flour beetle, *Tribolium castaneum* (Coleoptera: Tenebrionidae). Pestic Biochem. Physiol. 179: 104941.

Ozdemir, E. and U. Gozel. 2017. Efficiency of some plant essential oils on root-knot nematode *Meloidogyne incognita*. J. Agric. Sci. Technol. 7: 178–183.

Ozdemir, E. and U. Gozel. 2018. Nematicidal activities of essential oils against *Meloidogyne incognita* on tomato plant. Fresenius. Environ. Bull. 27:4511–4517.

Parra Amin, J.E., L.E. Cuca and A. González-Coloma. 2021. Antifungal and phytotoxic activity of benzoic acid derivatives from inflorescences of *Piper cumanense*. Nat. Prod. Res. 35: 2763–2771.

Patidar, R.K., D. Sen, M. Pathak and R.C. Shakywar. 2016. Effect of essential oils on mortality, hatching and multiplication of root-knot nematode, *Meloidogyne incognita* and its impact on plant growth parameters. Int. J. Agric. Environ. Biotechnol. 9: 887–895.

Peter, R., M.E. Josende, J. da Silva Barreto, D.G. da Costa Silva, C.E. da Rosa and F.E. Maciel. 2022. Effect of *Illicium verum* (hook) essential oil on cholinesterase and locomotor activity of *Alphitobius diaperinus* (panzer). Pestic. Biochem. Physiol. 181: 105027.

Prabha, S., A. Yadav, A. Kumar, A. Yadav, H.K. Yadav, S. Kumar, et al. 2016. Biopesticides—An alternative and eco-friendly source for the control of pests in agricultural crops. Plant. Arch. 16: 902–906.

Price, D.N. and M.S. Berry. 2006. Comparison of effects of octopamine and insecticidal essential oils on activity in the nerve cord, foregut, and dorsal unpaired median neurons of cockroaches. J. Insect. Physiol. 52: 309–319.

Priestley, C.M., E.M. Williamson, K.A. Wafford and D.B. Sattelle. 2003. Thymol, a constituent of thyme essential oil, is a positive allosteric modulator of human GABA(A) receptors and a homo-oligomeric GABA receptor from *Drosophila melanogaster*. Br. J. Pharmacol. 140: 1363–1372.

Putievsky, E., N. Dudai and U. Ravid. 1997. Cultivation, selection, and conservation of oregano species in Israel. Proc. IPGRI Internat. Workshop on Oregano: 102–109.

Raveau, R., J. Fontaine and S. and A. Lounès-Hadj Sahraoui. 2020. Essential oils as potential alternative biocontrol products against plant pathogens and weeds: A review. Foods. 21(9): 365.

Ravensberg, W.J. 2015. Commercialisation of microbes: Present situation and future prospects. pp. 309–317. *In:* B. Lugtenberg (ed.). Principles of Plant-Microbe Interactions. Microbes for Sustainable Agriculture. Springer International Publishing Switzerland.

Renoz, F., S. Demeter, H. Degand, S.C. Nicolis, O. Lebbe, H. Martin, et al. 2022. The modes of action of *Mentha arvensis* essential oil on the granary weevil *Sitophilus granarius* revealed by a label-free quantitative proteomic analysis. J. Pest. Sci. 95: 381–395.

Reuveni, M., E. Sanches and M. Barbier. 2020. Curative and suppressive activities of essential tea tree oil against fungal plant pathogens. Agronomy. 10.

Rienth, M., J. Crovadore, S. Ghaffari and F. Lefort. 2019. Oregano essential oil vapour prevents *Plasmopara viticola* infection in grapevine (*Vitis vinifera*) and primes plant immunity mechanisms. PLoS One. 14: 1–29.

Sarkhosh, A., B. Schaffer, A.I. Vargas, A.J. Palmateer, P. Lopez, A. Soleymani, et al. 2018. Antifungal activity of five plant-extracted essential oils against anthracnose in papaya fruit. Biol. Agric. Hortic. 34: 18–26.

Schippmann, U., D. Leama and A. Cunningham. 2006. A comparison of cultivation and wild collection of medicinal and aromatic plants under sustainability aspects. pp. 75–95. *In:* R.J. Bogers, L.E. Craker and D. Lange (eds.). Medicinal and Aromatic Plants. Springer. The Netherlands.

Scorzoni, L., A.C.A. de Paula e Silva, C.M. Marcos, P.A. Assato, W.C.M.A. de Melo, H.C. de Oliveira, et al. 2017. Antifungal therapy: New advances in the understanding and treatment of mycosis. Front Microbiol. 8.

Sellamuthu, P.S., D. Sivakumar, P. Soundy and L. Korsten. 2013. Essential oil vapours suppress the development of anthracnose and enhance defence related and antioxidant enzyme activities in avocado fruit. Postharvest. Biol. Technol. 81: 66–72.

Shahina, Z., A.M. El-Ganiny, J. Minion, M. Whiteway, T. Sultana, et al. 2018. *Cinnamomum zeylanicum* bark essential oil induces cell wall remodelling and spindle defects in *Candida albicans*. Fungal Biol. Biotechnol. 5: 3.

Soudani, S., C. Poza-Carrión, N. De la Cruz Gómez, A. González-Coloma, M.F. Andrés and M. Berrocal-Lobo. 2022 Essential oils prime epigenetic and metabolomic changes in tomato defense against *Fusarium oxysporum*. Front. Plant Sci. 13: 804104. doi: 10.3389/fpls.2022.804104.

Spadaro, D., J. Herforth-Rahmé and J. van der Wolf. 2017. Organic seed treatments of vegetables to prevent seedborne diseases. Acta. Hortic. 1164: 23–32.

Sukegawa, S., K. Shiojiri, T. Higami, S. Suzuki and G.-i. Arimura. 2018. Pest management using mint volatiles to elicit resistance in soy: mechanism and application potential. Plant. J. 96: 910–920.

Tak, J.H. and M.B. Isman. 2015. Enhanced cuticular penetration as the mechanism for synergy of insecticidal constituents of rosemary essential oil in *Trichoplusia ni*. Sci. Rep. 5: 12690.

Tak, J.H. and M.B. Isman. 2016. Metabolism of citral, the major constituent of lemongrass oil, in the cabbage looper, *Tricohoplusia ni*, and effects of enzyme inhibitors on toxicity and metabolism. Pestic Biochem. Physiol. 133: 20–25.

Tak, J.H. and M.B. Isman. 2017. Acaricidal and repellent activity of plant essential oil-derived terpenes and the effect of binary mixtures against *Tetranycus urticae* Koch (Acari: *Tetranychidae*). Ind. Crops. Prod. 108: 786–792.

Tak, J.H., E. Jovel and M.B. Isman. 2017. Effects of rosemary, thyme and lemongrass oils and their major constituents on detoxifying enzyme activity and insecticidal activity in *Trichoplusia ni*. Pestic. Biochem. Physiol. 140: 9–16.

Tamokou, J.D.D., A.T. Mbaveng and V. Kuete. 2017. Antimicrobial activities of african medicinal spices and vegetables. pp. 207–237. *In:* V. Kuete (ed.). Medicinal Spices and Vegetables from Africa, Academic Press.

Tong, F. and J.R. Coats. 2010. Effects of monoterpenoid insecticides on [^{3}H]-TBOB binding in house fly GABA receptor and ^{36}Cl-uptake in American cockroach ventral nerve cord. Pestic. Biochem. Physiol. 98: 317–324.

Tong, F., A.D. Gross, M.C. Dolan and J.R. Coats. 2013. The phenolic monoterpenoid carvacrol inhibits the binding of nicotine to the house fly *nicotinic acetylcholine* receptor. Pest. Manag. Sci. 69: 775–780.

Turek, C. and F.C. Stintzing. 2012. Impact of different storage conditions on the quality of selected essential oils. Food Res. Int. 46: 341–353.

Turek, C. and F.C. Stintzing. 2013. Stability of essential oils: A review. Compr. Rev. Food. Sci. Food. Saf. 12: 40–53.

Vilaplana, R., L. Pazmiño and S.Valencia-Chamorro. 2018. Control of anthracnose, caused by *Colletotrichum musae*, on postharvest organic banana by thyme oil. Postharvest. Biol. Technol. 138: 56–63.

Vining, K.J., K.E. Hummer, N.V. Bassil, B.M. Lange, C.K. Khoury, et al. 2020. Crop wild relatives as germplasm resource for cultivar improvement in mint (*Mentha* L.). Front. Plant Sci. 11: 1217. doi: 10.3389/fpls.2020.01217.

Werrie, P.Y., B. Durenne, P. Delaplace and M.L. Fauconnier. 2020. Phytotoxicity of essential oils: opportunities and constraints for the development of biopesticides. A Review. Foods. 14: 1291.

World Health Organization. 2003. WHO guidelines on good agricultural and collection practices (GACP) for medicinal plants. World Health Organization.

Wu, X.J., T.S tahl, Y. Hu, F. Kassie and V. Mersch-Sundermann. 2006. The production of reactive oxygen species and the mitochondrial membrane potential are modulated during onion oil-induced cell cycle arrest and apoptosis in A549 cells. J. Nutr. 136: 608–613.

Yousif, L, O. Belmehdi, B. Abdelhakim, N. Skali Senhaji and J. Abrini, 2021. Does the domestication of *Origanum compactum* (Benth) affect its chemical composition and antibacterial activity? Flav. Fragr. J. 36: 264– 271.

8

Production, Market and Bioeconomy of *Vetiveria zizanioides* L. (Poaceae) Essential Oil

Haroldo A. Silva[1], *Suiane S. Santos*[1], *Sthephanie S.S. Tavares*[1], *Josiane E.A. Silva*[1], *Subhash Chander*[2], *Amit C. Kharkwal*[3], *Rajendra Awasthi*[4], *Neerupma Dhiman*[5], *Bhupesh Sharma*[5], *Harsha Kharkwal*[2] and *Valdir F. Veiga-Junior*[1*]

The Importance of Vetiver Essential Oils in the Industrial Scene

There are several biotic and abiotic factors that can affect plants and modify their growing pattern, the biomass production and, also the specialized metabolites production, such as volatile compounds. The effects in essential oils can be easily observed, since they are very sensitive even in yield as in the chemical composition, being strongly affected under different growing conditions, such as organic and inorganic fertilizers; the presence of elicitors, such as herbivores; the amount of insolation, and so on. As several plants produce commercially useful essential oils, studies on how to improve their yield and potentialize the production of target metabolites is considered a very important research area. *Vetiveria zizanioides* (L.) is a perennial grassy herb belonging to the Poaceae family, originally from India, but now is widely distributed in tropical and subtropical regions. The essential oil obtained from the roots presents applications in the pharmaceutical industries, herbal formulations, and in perfumery, in the composition of major international perfume brands. This chapter broadly

1 Military Institute of Engineering, Rio de Janeiro, Brazil.
2 Amity Institue of Phytochemistry and Phytomedicine, AUUP, Noida, India.
3 Amity Institute of Microbial Technology, AUUP, Noida, India.
4 Department of Pharmaceutical Sciences, School of Health Sciences & Technology, UPES University, Dehradun, India.
5 Amity Institute of Pharmacy, AUUP, Noida, India.
* Corresponding author: valdir.veiga@gmail.com

investigates the different aspects of this essential oil, starting with some of the most important studies on the optimization of its yield and composition.

Commercial Production of Vetiver Essential Oil and its Market Value

The material preparation for procuring vetiver oil by steam-distillation from *V. zizanioides* roots is not easy. Digging and lifting of roots, and their cleaning requires extreme efforts (Lavania 2003a). Due to these reasons, it is difficult to set up a vetiver oil industry in new and untried locations (Robbins 1982). Vetiverol is the principal alcoholic component of *V. zizanioides* which is used as a perfumery ingredient (Belhassen et al. 2015, Shabbir et al. 2019). Vetiverol is an intermediate used in the synthesis of vetiveryl acetate. The acetylated derivative of vetiverol (vetiver acetate) can be also used rather than its principal alcoholic form (vetiverol). The odour of acetylated derivatives is milder than that of the base oil. It is more expensive and used in the synthesis of higher-class products. The synthesis of acetylated derivatives requires intricate unit operations and therefore the production of these derivatives is limited to only a few companies involved in the use of these derivatives (Robbins 1982).

Import and Export of Vetiver on the World Stage

Trading Scene of Vetiver Essential Oil in France

France is a major importer of vetiver oil with around 50 tons consumption in the perfume industries. The trade of vetiver oil in France is mainly handled by firms in Marseille and Bordeaux. Half of the total vetiver oil imported in France is re-exported in unprocessed or re-distilled and refined form. Haitian oil is the most used oil, which is about three-fold more than other vetiver essential oils in France. France is also a regular importer of vetiver oil from Brazil and China. But the Brazilian and Chinese oil are not as popular as the Haitian oil. The use of Bourbon oil is declining in the French perfumery industries due to the increased popularity of vetiver oil (Robbins 1982, Skaria 2007).

Trading Scene of Vetiver Essential Oil in the United Kingdom

The import of vetiver oil in the United Kingdom is mainly handled by several traders, but few larger users are now directly trading with the exporters. The annual import of vetiver oil in the United Kingdom is more than 20 tons/year. It is used in both forms, either as pure oil or its derivatives like vetiver acetate mainly to provide a fragrance for cosmetic products, aerosols, soaps, and household products. The major fraction of the oil imported from France and Indonesia is consumed domestically. Haitian oil obtained from the US dealers is less popular in the United Kingdom. Despite the good reputation of Bourbon oil, its demand has declined in the USA due to the high prices. Bourbon and Haitian oils, and their derivatives are generally used in more expensive applications. Chinese oil received comparatively more popularity due to its special character and easy supply (Robbins 1982).

Trading Scenario of Vetiver Essential Oil in the United States

The United States is the largest consumer of vetiver oil with an estimated annual trade of >100 tons/year. It is extensively used within the United States. Haiti is the main source of vetiver oil in US imports. However, the import of Indonesian vetiver has also increased markedly during the 1970s. Unprocessed Haitian and Indonesian oils received equal status within the United States users. Indonesian oil is the choice of oil to produce vetiver acetate and vetiveryl acetate derivatives. Comparatively more fair quality Brazilian oil has a very low trade value in the United States. The popularity of Bourbon vetiver oil has declined in the USA during recent years. However, it is still preferred for perfumery applications. Vetiver oil imported from Indonesia is mainly handled by traders from New York. Soap manufacturers are the primary customers of unprocessed oil. However, the acetylated derivative of vetiverol (vetiver acetate) is also gaining popularity. During the 1970s, the photosensitization associated harmful effects of vetiver oil on the human skin ware scuttlebutt. However, such allegations were never proven, but still, they had impacted the overall trade of oil in the United States (Robbins 1982).

Trading Scene of Vetiver Essential Oil in Switzerland

Perfume and flavor producers for the European market are the major consumers of vetiver oil in Switzerland. There is no documentary evidence of vetiver oil statistics is available for Switzerland. However, an estimated import of vetiver oil in Switzerland is more than 30 tons. Indonesian oil is the most used vetiver oil in Switzerland, followed by the Chinese and Haitian oils. The trade of Bourbon oil in Switzerland is in very low quantities. Despite acidity issues of Chinese oil, its consumption is relatively large (about 5 tons/year). It is mixed with Indonesian oil and the oil is acetylated before its use. The oil with an adequate content of vetiverol and a good odour is the choice of buyers. Generally, Swiss buyers purchase vetiver oil directly from the source. However, the trade of Haitian oil in Switzerland is carried out through dealers of North America and/or Western Europe (Robbins 1982).

Trading Scene of Vetiver Essential Oil in the Netherlands

Vetiver oil has gained a good reputation among users in the Netherlands because of fashion movements and issues with synthetics. The use of vetiver oil in the Netherlands is slowly increasing. The average import of vetiver oil in the Netherlands is about 5 tons/year. The use of vetiver oil in the Netherlands is limited to only large processing and compounding companies. Haitian vetiver oil is mainly supplied by the United States to the Netherlands. The demand for Indonesian oil is extremely small in the Netherlands. High-quality oil from France is also purchased by the buyers of the Netherlands. A larger fraction of the oil consumed is converted to vetiveryl acetate or vetiver acetate before its use. The buyers from Netherlands of vetiver oil have direct contacts with overseas suppliers. However, a small portion of the trade is carried out through the traders in Rotterdam and Amsterdam (Robbins 1982).

Trading Scene of Vetiver Essential Oil in the Federal Republic of Germany

West Germany is a small trader of vetiver oil. However, the imports of vetiver oil-based ready-made perfume compounds into West Germany is at a greater level (approximately 6 tons/year). This is mainly imported in the form of expensive redistilled and refined oils from France. The trade of vetiver oil in West Germany is generally handled by the dealers in Bremen and Hamburg. This shows that the imports of vetiver oil in West Germany is not directly from its main source countries like Indonesia and Haiti. Haitian oil is mainly used as a source of vetiver acetate. Indonesian oil is also used as a cheap source of vetiver acetate. Favored Reunion vetiver oil (Bourbon oil), used directly in its own right, is mainly imported from France. Vetiver oil is used by various cosmetic industries of West Germany as a flavoring material in men's lotions, deodorants, and soaps (Robbins 1982).

Main Production Centers of Vetiver Essential Oil

Brazil, China, Haiti, Indonesia, and Reunion Islands are the main sources of vetiver oil (Robbins 1982, Shabbir et al. 2019). Decades back, Guatemala (a country in Central America) was an exporter of vetiver oil. However, Guatemala does not produce the oil now (Sreenath et al. 1994). India is also the producer of vetiver oil, but never gained appreciable international recognition. South and Central America, and Indian Ocean also produce vetiver oil but, the oil production in these parts is not sustained. The annual world production of vetiver oil is 245–265 tons (although no reliable published statistics are available) (Robbins 1982). The present global demand of vetiver oil is about 500 metric ton and approximately 400% growth in essential oil production has been recorded in recent years (Shabbir et al. 2019).

The production of vetiver essential oil has followed with the growing world demand, although a notable variation is found in the main producing regions over the decades. Reunion Islands, for example, have suffered a drastic decrease in its annual vetiver essential oil production over the decades. In 1960, the island produced 50 tons/year (25% of the world's production) to only 6 tons/year (2.4%) in 1989 (Guadarrama 1990).

Between 1970 and the early 1980s the world production of vetiver oil reached between 245–265 tons and world exports ranged between 235–255 tons. The main suppliers of vetiver oil were Haiti and Indonesia (Java), the biggest producers, with an average production of 100 tons/year (Robbins 1982). The Réunion Islands, Brazil, China and, to a lesser extent, India also had a significant production.

The world market for vetiver oil in the late 1990s grew to approximately 400 tons/year. The average return per hectare of oil was approximately 40 kg/hectare and the duty-free price in the European Union was approximately 110 ECU/kg. In the following decade, in mid-2000s, the world trade of essential oil decreased compared to the previous period and maintained its world production at approximately 250 tons. The main suppliers were again Haiti, India, Japan, Brazil, China, Java and Reunion Islands (Maffei 2002, Robert 2009).

In the following decades from the year 2000 onwards a lack data on the world production of this essential oil, made it difficult to monitor the variations in production in countries in the 21st century. The most recent data found showed that Haiti reached its highest production peak and was considered the world's largest exporter in 2015, generating US$ 15.7 million per year with vetiver oil (Oscar 2015, INC. 2015).

Vetiver Essential Oil Production in Brazil

Brazil is the major producer of high-quality vetiver oil (Lavania 2003a), it is considered better than the oil produced in China and Indonesia (Barbieri and Borsotto 2018). Due to the rigorous control of the oil import in Brazil and growing demand for cosmetics, a considerable portion of this oil is used by the internal market of Brazil. Only a small portion of the vetiver oil produced in Brazil (<10 tons/year) reaches the international market (Robbins 1982, Lavania, 2003a, Martinez et al. 2004). In 2005, the volume of the vetiver oil exported reached 5.2 tons, leading the market and increasing its export in consecutive years, greater than 2 tons. In the last 10 yr (2012–2021), Brazil never exported this amount again (Figure 8.1).

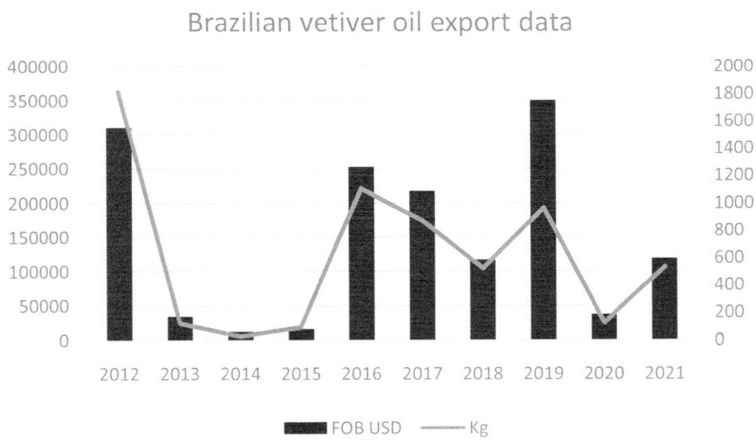

Figure 8.1 Vetiver oil production in Brazil.

In the economic scenario, Brazil is an important producer and consumer of vetiver essential oil, with a high demand from the cosmetics and perfume industries. Nationally, the largest producers of the species and, consequently, of vetiver oil are the states of Rio Grande do Norte, São Paulo, and Rio de Janeiro. The main buyers of the Brazilian vetiver essential oil are the United States, United Kingdom, Spain and France (Martinez et al. 2004).

Vetiver Essential Oil Production in China

China has been involved in the export of vetiver essential oil since the 1970s. The vetiver oil production level of China has been estimated as about 25 tons/ year.

China exports about three-quarters of the total oil produced (Maffei 2002). The oil produced in Kwangtung Province is considerably better in quality than the oil produced in the Shanghai Province. However, the Chinese vetiver oil did not get considerable interest of Western users. The acid content of Chinese vetiver oil is high, this led to a low content of vetiverol and its acetate derive. Due to this reason, the cost of Chinese vetiver oil is less than the oil produced by other countries. However, despite its low yield, it has a limited demand for the isolation of vetiver acetate and blending purposes. The average annual import of Chinese vetiver oil into the Western industrialized countries is below the estimated level of Chinese exports than the oil produced by Indonesia, Haiti, or Reunion (Robbins 1982).

Vetiver Essential Oil Production in Haiti

Haiti is the best large volume vetiver oil producer (Martinez et al. 2004). However, the quality is not as good as the Reunion oil (Bourbon oil). The essential oil market has been dominated by the Haitian vetiver oil. The annual Haitian vetiver oil production is about 100 tons that is all exported. Haitian vetiver oil contains >50% vetiverol. However, this amount is lower than the content present in the Reunion oil. Haitian oil is suitable to produce vetiveryl acetate or vetiver acetate. The strongly centralized marketing structure of the Haitian industry and the relatively limited number of dealers in the importing countries have limited the Haitian vetiver oil trade. A linear decrease in Haitian vetiver oil trade and prices was recorded during the 1970s due to the high officially fixed prices, periodic communication issues, and reported lapses in standard trading procedures. Presently, less rigid marketing practices and the communications problems in Haitian industry are expected to limit the trade loss (Robbins 1982).

Vetiver Essential Oil Production in Indonesia

The annual production of Indonesian (Java Island) vetiver oil is around 100 tons/year. Indonesian vetiver oil is the second important oil in the trade followed by Haitian oil. Despite the increased demand of Indonesian oil, its quality is variable due to the variation in the care taken during the distillation process. This is mainly because of its ruling and very substantial price fluctuations. Low vetiverol content, burnt odor, high acidity, and the presence of contaminants are the major uneasy factors for users. The Indonesian oil has been reported to have a high acid number and low vetiverol contents (47–48%) due to improper distillation, premature harvesting, and less care during the processing, probably related to the rising costs in conjunction and cash problems. Despite the Indonesian oil not gaining the reputation as of Haitian oil, still like the Haitian oil, the oil and derivatives of Indonesian oil are used in a range of products at a lower price. In recent years, Indonesian vetiver oil has gained popularity in France and the USA. The regular production of Indonesian vetiver oil depends on the international market price. During the beginning of the 1980s, the Indonesian vetiver oil production was minimum due to the release of Haitian stockpile in the international market.

However, the prices of Indonesian vetiver oil recovered with increased attention of its regular importers during early 1982 (Robbins 1982).

Vetiver Essential Oil Production in the Reunion Islands

Locally distilled Reunion vetiver oil is also known as Bourbon vetiver oil. This is the best-known vetiver oil in terms of odor and vetiverol contents (60%). Proper distillation and care during harvesting and processing lead to a limited production rate. The price of the oil is higher than the vetiver oils obtained from other sources. Reunion vetiver oil has a better quality and the high price, despite it also has an official price-fixing like Haitian vetiver. During past decades, a decline in the production rate of Reunion vetiver oil has been recorded because of its high price. The annual production of Reunion vetiver oil is about 20 tons. The oil producers need economic support for replacing aging equipment and installing new equipment. The labor cost has also been raised. Despite these limitations, the quality is still very good (Robbins 1982).

Vetiver Essential Oil Production in Other Countries

Angola was a regular international supplier of vetiver oil till the early 1970s. The supply of Angola vetiver essential oil reached zero during the Civil War of Angola and no recovery was recorded in the oil production. Zaire was also a small producer of vetiver oil under Belgian management which had declined to zero by the early 1970s. Presently the oil production in Zaire has been closed. India produces vetiver oil on a very small scale and its production is also irregular. The grade of Indian vetiver oil is average, and price is also high. Guatemala was the exporter of about 5 tons of vetiver oil annually until the early 1970s. Presently Guatemala does not produce vetiver essential oil. Malaysia produces a very small amount of vetiver oil which is consumed locally. There is no international market evidence of Malaysian vetiver oil. The Republic of South Africa is another small producer of vetiver oil. South African vetiver oil is of very high quality and close to that of Reunion vetiver oil (Robbins 1982).

Technologies Used in Prospecting for Products Based on Vetiver Essential Oils

Phytochemical Studies on Vetiver Essential Oil

Many volatile phytoconstituents have been isolated by different researchers from vetiver oil. The types of constituents and their percentage content are found to be dependent on certain factors like type of species, geographical location, environmental conditions, soil conditions, seasonal variation, and methods of distillation. Few constituents have a wide distribution in most vetiver species and the individually reported for certain pharmacological activities are; Khusimol also known as zizanol, Isonootkatol also called vetiselinenol, α-vetivone (isonootkatone), β-vetivone cyclocopacamphan-12-ol (epimer A), β-vetivenene, β-vetispirene, γ-vetivenene, α-cadinol, β-eudesmol, khusenic acid, α-amorphene and b-calacorene (Pripdeevech et al. 2006, David et al 2019, Suyono and Susanti 2020).

Khusimol (zizanol) is an alcohol, first isolated from the high boiling fractions of vetiver oil. It has reported for inhibition of vasopressin and for anti-tubercular activity against some drug resistant strains (Rao et al. 1994, Dwivedi et al. 2013).

Vetiselinenol (isonootkatol) is reported for antifungal effects against *Microsporum canis, Microsporum gypseum, Trichophyton mentagrophytes* and *Trichophyton rubrum* with synergistic effects with some other constituents from vetiver oil. The compound possesses anti-anti-bacterial activity against some pathogenic bacterial strains and fungal species (Rios and Recio 2005, Wagner and Ulrich-Merzenich 2009, Santos et al. 2014).

Constituents like Beta-vetivenene, alpha and beta vetivone showed potent anti-oxidant activity in some studies (Kim et al. 2005). Moreover, Beta-vetivene also showed relief in the GIT related discomfort (Virshette et al. 2020). Some constituents like isokhusenic acid, alpha-vetivone, Nootkatone, dehydroaromadendrene, and isolongifolene showed a prominent nephroprotective effect against doxorubicin-induced nephrotoxicity (Amarasiri et al. 2021).

A constituent nootkatone and its essential oil exhibited repellence and antifeedant activity against *Coptotermes formosanus* termite, which causes considerable damage to wooden items and trees. (Chen et al. 2003, Anyango et al. 2019). A study also demonstrated the insect-repellent activity of eight valencenoids namely; valencene tetrahydronootkatone, isonootkatone (α-vetivone), nootkatone, 11,12-dihydronootkatone, tetrahydrovalencene nootkatol and 1,10-dihydronootkatone against the *Formosan subterranean* termites (Zhu et al. 2003).

Although, a significantly large number of pharmacological studies have been performed using the essential oil or extracts of vetiver, but relatively limited pharmacological studies are conducted on individual constituents of vetiver oil. Future studies on individual compounds may reveal their more pharmacological potential.

Biological Activities Associated with Vetiver Essential Oil

Vetiver has been used for various medicinal and aromatic benefits since centuries, particularly in countries like India, Thailand, Pakistan, Senegal, Sri Lanka, Mauritius, and Indonesia. Furthermore, the oil have been used to treat other ailments like mouth ulcers, urinary tract infections, epilepsy, burns, snake bites, fever, headaches, cardiac related disorders, ailments related to the liver, vomiting and GIT related disorder (Gurib-Fakim et al. 1993, Khan et al. 1997, Sastry 1998, Shealy 1998, Lingga 2001, Pripdeevech et al. 2006, David et al. 2019). Diverse types of pharmacological activities have been reported of the vetiver plant such as antibacterial, hepatoprotective, antifungal, antitubercular, nephroprotective anti hyperglycemic, and antidepressant (David et al. 2019, Amarasiri et al. 2021). Some studies also marked their anti-cancer activity. (Powers et al. 2018), hepatoprotective action against ethanol intoxication (Parmar et al. 2013), antioxidant and anti-inflammatory activity (Grover et al. 2021).

Main Commercialized Bioproducts Based on Vetiver Essential Oil

Vetiver essential oil and its derivatives have many olfactory facets and their unique organoleptic characteristics are used in fine perfumery in the preparation of male and female perfumes and in eau de toilette. Among the classics are the perfumes Channel No. 5, Azzaro and Guerlain toilet water. In the preparation of perfumes, the essential oil can be used as a solid base over which other fragrance notes are superimposed and occasionally contributes mainly to a fragrance. Vetiver essential oil is also present in various cosmetic applications such as soaps, skin lotions, deodorants (Maffei 2002).

In the field of food science, vetiver oil is an edible oil considered valuable in the traditional Chinese market and is officially recognized as a safe natural food additive. In India it is used as a food additive in ice creams and in the manufacture of fresh products, syrups, beverages and for food preservation. Vetiver oil can be combined with many commercially known essential oils such as chamomile, ginger, jasmine, lavender, lemongrass, patchouli, sandalwood, and vanilla which broadens its application as a food flavoring. Due to its antimicrobial activities proven in several studies, this essential oil has great potential in food preservation (Chou et al. 2016).

Sustainable Management of Vetiver for the Use of its Plant Derivatives

Agro-industrial Vetiver Cultivation Models

The vetiver plant in cultivation rarely produces seeds. The methods of propagation are: Tiller is the most popular part of the vetiver plant used in propagation since it is available in large quantities, uses a simple technique, and gives good results. The culm of the vetiver grass is strong, hard, and lignified, having prominent nodes with lateral buds that can form roots and shoots on exposure to moist conditions.

Laying the cut pieces of culms on moist sand, or better under a mist spray, results in the rapid formation of roots and shoots at each node. A clump is formed when a plant has been grown for a certain period and produces numerous tillers in all directions. Unlimited number of plantlets can be produced in a septic condition from the explants derived from the shoot tip, lateral bud, young inflorescence, etc. On attaining a good size, these 'plantlets' can be transplanted in containers or in fields like tillers, although much smaller in size. Tissue-cultured plantlets can be produced within a relatively short time with reasonable costs (Liu et al. 2021).

They also have certain advantages over other planting materials in that they are small in size, easy to transport, and free from pathogens (as they are grown, and still remain, in an aseptic condition) which makes them safe for international movement, especially across the countries with strict plant quarantine system (Maffei 2002).

Ecophysiological Characteristics of Agro-Industrial Vetiver Production

Vetiver can be grown on almost every kind of soil (Smitha et al. 2014). However, light soils, should be avoided as the roots grown in this soil produce a very low percentage of oil. Well drained sandy loam and red lateritic soils rich in organic matter are considered to be ideal for cultivation. It can also be cultivated in clay loam soil, but it is better to avoid clayey soil. It can be grown in a wide pH range even in saline and alkaline soils with a pH of 8.5 to 10. A flat site is acceptable, but watering must be monitored to avoid water logging, that will stunt the growth of young plantlets. Mature vetiver, however, thrives under waterlogged conditions.

Vetiver's unique morphological, physiological, and ecological characteristics, including its tolerance to highly adverse conditions, has played a key role in the area of environmental protection and land rehabilitation. These include tolerance to the following adverse conditions: (i) acidity, manganese, and aluminium, (ii) salinity and sodicity, and (iii) heavy metals like arsenic, cadmium, copper, chromium, lead, mercury, nickel, selenium, and zinc.

Vetiver is tolerant to a wide range of temperature ranging from –15°C to +55°C, depending on growing region (Smitha et al. 2014). The optimal soil temperature for the growth of roots is 25°C. Root dormancy occurs when temperature goes below 5°C. Under frosty conditions, shoots become dormant and purple, or even die, but the underground growing points survive and can regrow quickly if conditions improve. Shading affects the vetiver's growth, but partial shading is acceptable. It is tolerant to drought, flood, submergence and grows luxuriantly in places having a moderately humid climate with annual rainfall of 1000 to 2000 mm. It can also be grown as an irrigated crop in other suitable places with scanty rainfall.

Vetiver needs bright sun with high temperatures but cannot stand low temperatures (Smitha et al. 2014). It is a tropical grass which can well adapt to different environmental conditions. When planted on contour lines on slopes, the clump which stands above the ground will produce tillers, forming a green hedge. This makes it capable of trapping crop residues and silts eroded by runoff, leading to natural earth terrace formation. With a deep, dense root system spreading vertically, rather than horizontally, vetiver can be tolerantly grown under adverse conditions.

Not only vetiver can tolerate adverse soil conditions, but it can absorb toxic substances like pollutants, pesticides, and heavy metals into its biomass (mostly in its extensive roots), thereby diluting such toxic substances in the soils and water, making them safer in agricultural and non-agricultural activities (Smitha et al. 2014). One such activity is the use of vetiver to purify eutrophicated water in lakes, leachate from garbage landfills, etc. It can also be used to absorb heavy metals from quarries, and such valuable metals can be recycled by extraction from the vetiver biomass.

Sustainable Cultivation of Vetiver for the Production of Organic Plant Derivatives

The application of vetiver grass as a technology for soil and water conservation was first developed in India by the World Bank in the 1980s (Truong 2000). Some of the applications of vetiver grass technology, which could be summed up to the enhancement of sustainable agricultural development, including its use for soil erosion and sediment control on sloping farmlands and floodplains (Truong 1993, Oshunsanya 2013a, Oshunsanya 2013b, Oshunsanya et al. 2014); rehabilitation of saline and acid sulphate soils (Truong and Baker 1998); bioremediation of agrochemicals (Pinthong et al. 1998, Oshunsanya and Ewetola 2011); biological pest control (Chen 1999); and on- and off-site heavy metal pollution control (Truong and Baker 1998, Oshunsanya and Ewetola 2011) amongst others.

From early periods, vetiver oil has been produced locally and used as perfumes and medicine. However, only in the recent decades they are produced and marketed in the industrial scale. Vetiver fibrous roots contain, in average, 1–3% (dry weight basis) of oil. The crop oil recovery yield is reported as 10–30 kg of oil per hectare. Vetiver oil is largely produced worldwide, coming mainly from Haiti, Indonesia, and India, with an estimated capacity of 300–350 tons/year (Lal 2013). The essential oil is usually obtained by steam distillation, leaving abundant root residues (Altenor et al 2013). Vetiver oil is greatly appreciated for the formulation of high-grade perfumes due to its woody, earthy and grapefruit odor. Vetiver oil is mainly used for its dry woody base notes in perfumes and 250 kg of vetiver root are used for producing 1 kg oil extract (Sarrazin, 2017).

Apart from its value in the perfumery industry, vetiver oil is also used in cosmetics since the olden times. The vetiver roots are soaked in coconut hair oil to give fragrance and is still practiced in some rural parts of southern India. In recent years, the use of the oil is widespread in several cosmetic products such as soaps, deodorants, lotions, shampoos, and skin bleach. Vetiver oil is suitable for cosmetics due to its fragrance, low volatile additive characters, and safety specifications (Moussou et al. 2016). Vetiver oil use in cosmetics is expected to increase in the future as consumers are demanding more sustainable cosmetics with natural ingredients (Gupta et al. 2015).

Like its application in perfumes and cosmetics, vetiver oil is used as a medicine since ancient times to treat dermatitis, hemorrhoids, fever, rheumatism, neuralgia, and fungal growth control (Porras and Jochamowitz 2002, Balasankar et al. 2013, Smitha et al. 2014).

In recent years, their application in medicine is supported by several research findings. Porras and Jochamowitz (2002) received a patent for the invention relating to vetiver grass extract for increasing or restoring hair growth or preventing hair loss. Another research study data suggests that vetiver grass oil has therapeutic potential for both cosmetic and metabolic health care products through bioactivity tests in a human skin disease model. Vetiver oil showed significant anti-proliferative activity, inhibited collagen III production, strongly modulated global gene expression and impacted several critical physiological processes signaling pathways related to tissue remodeling and cholesterol metabolism (Han and Parker 2017).

The effect of vetiver oil on tuberculosis was studied by Saikia et al. (2012) who concluded that vetiver oil was active against both virulent and avirulent strains of *M. tuberculosis*. The vetiver root extract could be potentially used as an antituberculosis agent. Moreover, vetiver oil has been tested as a mosquito repellent along with other plant oils such as citronella, hairy basil, and catnip.

The results confirmed that vetiver oil could serve as potential mosquito repellent products against *Cx. quinquefasciatus* and *An. minimus* mosquitos. A similar study was reported for the malaria vector *Anopheles stephensi*, anti-inflammatory, anti-oxidant and anti/fungal property of vetiver oil (Lee and Lee 2010, Aarthi and Murugan 2012, Chou et al. 2012). Vetiver oil is also used in aromatherapy within hospice and palliative care for cancer patients, as it can alleviate pains caused by cancer and decrease symptoms like chronic fatigue, pain, insomnia and anxiety (Miller 2006).

Biotechnological Potential of Vetiver Plant Derivatives

By using new technology known as biotechnology, vetiver has been subjected to two kinds of studies, namely the tissue culture method of propagation and DNA finger printing to determine their variation. The first approach is to have rapid mass propagation using the explants derived from young shoots of young inflorescence and grow them in an aseptic condition to produce cell mass known as callus, which is then allowed to multiply until enough is produced (Eapen 2007). With the change of nutrient, callus differentiates into the root and shoot of the plantlet, which is still in an aseptic condition until reaching a good size before it is transferred to be grown in the nursery.

In the DNA-fingerprint analysis of genetic diversity of vetiver, it appears that only one *C. zizanioides,* genotype, 'Sunshine', accounts for almost all germplasms used in the soil and water conservation outside South Asia. Twenty-one accessions of vetiver (*Vetiveria zizanioides* (L.) Nash, sterile, oil type) and Khus (*V. zizanioides*, fertile, non-oil type) were analyzed using Random Amplified Polymorphic DNAs (RAPDs) by Adams et al. 2003. Nineteen of the accessions clustered strongly around the cultiven, Sunshine. Three accessions, Khus, Northern India, Kassel, Germany, and Guang Dong, China clustered loosely and were not closely related to the sterile oil-producing vetivers. Thirteen of the vetiver accessions were grown in test plots in Florida, the USA, Nepal, and Portugal. The largest growth was recorded in Nepal, followed closely by Florida and by the cooler, Mediterranean site in Portugal. No single genotype (DNA cultivar) grew best in every plot. The oil yields (% oil/dry wt.) were highest in Nepal and Portugal. Oil yields ranged from 0.29 to 9.61%. Essential oil production (g/plant) was highest in Nepal and Florida and ranged from 0.02 to 4.17 (g/plant).

Bioeconomy of Vetiver Essential Oil: From the Field to Industries

Vetiver as a Promoter of the Bioeconomy

The search for a new bioeconomic paradigm started in the 20th century, when there were changes in the environment through society and technology, which recognized the need to preserve natural resources. Bioeconomy entered politics and industry, promoting economic, social, and environmental development, seeking to renew several primary and secondary sectors. Among the sectors in which the bioeconomy is most active are the food, pharmaceutical, textile, forestry and chemical sectors (Dierz et al. 2018, Bruckner et al. 2019).

Bioeconomy uses mainly biotic resources, approaching agroecology and biotechnology, aiming at a linear economy in which the environmental impact related to industrial processes can decrease (D'Amato et al. 2020). Promising to mitigate the effects of climate change, proposing a renewable carbon source, the creation of businesses and new employment opportunities, especially in the rural sector (Ubando et al. 2020).

Vetiver essential oil plays an important role in bioeconomy, mainly in low-income regions, such as South/Central America (Haiti, Guatemala and Brazil) and Indic/Pacific Ocean (Malaysia, Java, and the Reunion Islands). This voluminous perennial herbaceous species, native from India, is tolerant to a wide range of stress conditions, such as salinity, acid, xerophyte and various soil conditions (Burger et al. 2017, Ishrath et al. 2019).

In 2003, it was estimated that about 250 tons of vetiver essential oil were traded worldwide, valued at US$20–200 million per year. In Haiti, the world's leading exporter, more than 50,000 families maintain their lives from the production of vetiver, in 2015 vetiver earned US$15.7 million per year for Haiti (Lavania 2003b, Burger et al. 2017).

Vetiver is highly valued for its production of essential oil which is one of the main economic markets in which the species is widespread. It can be extracted in a fertile form, produced in India and sterile without sowing in cultivated areas in other regions. This essential oil is widely used in the perfume industry producing modern female and male fragrances, and is also appreciated in modern aromatherapy for its sedative and skin-regulating qualities (Chomchalow and Chapman 2003, Lawrence 2016).

The production of vetiver is largely linked to the bioeconomy, which aims at economic development and technological innovations in the production and sale of products derived from biodiversity. Vetiver essential oil is part of the production of numerous perfumery products that are known worldwide, such as Miss Dior by Dior, Sr N vetiver by Natura, Pure vetiver TM by Azzaro, among others.

Sustainable Alternatives for the Use of Waste from the Vetiver Production Chain

The growing demand for vetiver essential oil has made the management in production even greater, with the main aim of producing plants with more morphophysiological quality and health, capable of meeting the planting objectives. One of the most striking morphological characteristics of this species is the fact that it does not produce rhizomes and despite producing inflorescences, its seeds are sterile. However, this feature allows its use in different biomes. As for the propagation methods of this grass, currently, the methods of separating shoots from the clump and by sectioning the stalk are used (Donjadee and Tingsanchali 2012, Chaves and Andrade 2013, Liu et al. 2021).

Leaves of *Vetiveria zizanioides* were analyzed for composition, showing: polysaccharides, hemicellulose (38%), cellulose (27%), 10% lignin content, which covalently binds to cellulose, in addition to 5% proteins. Another study showed: cellulose (32.6%), hemicellulose (31.5%) and lignin (17.3%) (Raman and

Gnansounou 2015). These factors contribute to the existence of so much residue being a resistant root. Vetiver ash makes up 3% of the total with a stiffness associated with silica present in 50% of that ash. Ash is a relevant part of the waste as it can be used for building construction items like mortar instead of cement according to studies (Nimityongskul et al. 2003) and is used to stabilize slopes. Fibers have been used in the manufacture of panels and sheets instead of wood, in addition to being used to manufacture mattresses in Ethiopia and for brake pads for automobiles in Brazil (Parisca and Tamayo 2006, Pereira 2006). Due to its composition of cellulose, hemicellulose, and lignin in the biomass of vetiver grass, it is considered a suitable raw material for producing bioethanol, and also in the production of fuel for generating steam and electricity.

Figure 8.2 shows the process that vetiver goes through until it reaches commercialization. It starts with the production of seedlings, with the formation of nurseries, using seedlings with bare roots or in plastic bags, later, the planting site must be chosen, with a deeper and sandy soil, which will help in the harvesting process and reduce the risk of flooding, planting should be done soon after the start of the rainy season, it is important that there is no shortage of water for better establishment of seedlings and to increase their survival rate (Chaves and Andrade 2013).

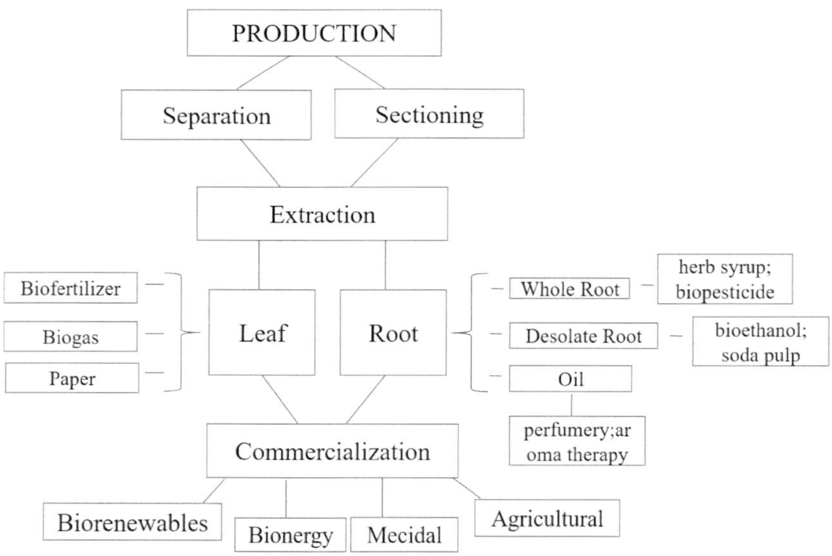

Figure 8.2 Vetiver production flowchart.

The lignocellulose residues generated by the vetiver industry can be used in the production of activated carbon (Gaspard et al. 2007) and to supply bioethanol production and biochemical productions in a biorefinery as it has a good performance. Its development, growing fast and in a relatively small planting area make it appealing. Vetiver powder is used in the production of furniture and household appliances in addition to other substances (Hirunpraditkoon and García 2009). Other residues are generated and have found use in several industries, whether solid (extracted material) or liquid (hydrolate) and can be inserted in various parts of the production chain, which explains its importance. The lack of data about residues is significant and the existing data tends to be old (Table 8.1).

Table 8.1 Industrial applications of vetiver grass

Industrial Application	Raw material	Product
Bioenergy	Leaf	Charcoal Biogas
Bioenergy	Leaf	Bioethanol
Biorenewables	Leaf	Vegetable oil Biolubrificant Paper
Agricultural	Leaf Root	Biopesticide Biofertilizer
Medical	Leaf Root Oil	Aroma therapy Anticancer Antioxidants

Source: Adapted from Raman et al. 2018.

The essential oil obtained from the leaves and roots of vetiver, is intended for various purposes within industries, such as the production of biofertilizers, bioethanol, paper, perfumery, among others.

The extraction of vetiver grass is usually done by three processing techniques whose main objective is to reduce environmental impacts, using high pressure gas extraction solvents, such as CO_2, which are characterized by the relative ease of separating the solvent, which makes the extracts purer. Therefore, the techniques of hydrodistillation, extraction with organic solvent and extraction with supercritical CO_2 are largely used to obtain the essential oil of vetiver (Raman et al. 2018).

Adding Value to By-products from Vetiver

Vetiver's industrial production is mostly focused on oil extraction as it is the base raw material for the HPPC (hygiene, perfumery, and cosmetics) and food industries that seek materials of natural origin. Vetiver grass is classified as having two types of roots, smooth or the secondary capillary, the secondary has in its composition non-polar compounds that are undesirable for the market, which makes chemical transformation of these compounds necessary (Lavania 2003c).

Another reason for chemical transformation is the prioritization of the essential oil by industries. Substances other than non-polar compounds are generated in the extraction of essential oil, by-products that can also be used via physicochemical transformation. Oil extraction is done by several processes such as: Hydrodistillation, extraction by organic solvents and extraction by supercritical fluid (Silveira et al. 2012).

For better extraction, some pre-treatments are necessary to improve the interaction of the oil and the extraction medium, which allows a better surface area for mass transfer by molecular diffusion. The first separation in the vetiver extraction in this pre-treatment separates the oil into two phases: the organic phase (extract) and the aqueous phase (refined). Figure 8.2 shows the extraction of vetiver with its products and by-products after the pre-treatments and treatments correctly chosen.

Among the by-products are carboxylic acids from the aqueous phase in the earlier figure. These acids are known not to contribute to the characteristic odor of vetiver essential oil and therefore need to go through new processes to add value. The study by Martinez et al. (2004) compared the total identified carboxylic acids in Brazilian vetiver oil with that from other sources and by the extraction treatment to understand the relevance of these by-products, with the neutral fraction of the oil being a part of interest. It is possible to verify that the amount of carboxylic acid can be high depending on the oil or extraction method, hence the need for treatment of the recovered acid.

Also in the aforementioned study, the chromatographic profile showed that from the amount of recovered acid, the major acid was zizanoic acid, sesquiterpene, with 76% presence (23% of the total extract), a substance mentioned for the first time in vetiver oil by Hanayama et al. (1973) together with epizizanoic acid (Table 8.2). Other lesser known acids in this oil are: khusenic acid isolated for the first time by Hanayama et al. (1973) from Indian vetiver oil (1957) and isokhusenic acid by Hanayama et al. (1973). Other studies cite nonaic acid (Pripdeevech et al. 2006) and hexadecanoic acid (Danh et al. 2010).

Table 8.2 Carboxylic acids by vetiver oil and methods

Vetiver Extraction—Quantities of Carboxylic Acids (%)	
Brazilian comercial oil	18.4
SFE extract	46.2
Hydrodistilled SFE extract	9.7
Neutral fraction from Brazilian oil	4.9
Hydrodistilled oil	34.4
Haiti oil	0.7
Bourbon oil	1.7
Java oil	4.7

For vetiver oil to be industrially valuable, it is necessary to have a composition that favors it with different odors, as in the case of sesquiterpenic alcohols, including the most striking alcohol, Khusimol (Leite 2012), in addition to eliminating all uninteresting by-products, that allow, for example, better miscibility in perfumes (Danh 2007). Carboxylic acids and certain hydrocarbons present in this oil tend to undergo a reaction for transformation or fractionation (Emishaw 2017), in the case of zizanoic acid, the conversion is focused on obtaining khusimol, the final product shown in Figure 8.2.

Khusimol is a central interest substance since it is considered a quality marker of vetiver oil for some producers (Talansier et al. 2008). Khusimol according to Champagnat et al. (2006) is one of the most characteristic components of the vetiver root, denoted as a fingerprint of this oil, as confirmed by Mallavarapu et al. (2012) by the quantity isolated experimentally.

Conclusion

Vetiver grass is a very suitable resource to be used in several industries, from bioenergy to perfumes and cosmetics. The world trade of its essential oil comes from Haiti, Brazil, China and the Reunion Islands to France and the USA. The oil chemical composition varies from each source, but some unique substances present high biotechnological interest.

Acknowledgment

This study was funded in part by the Fundação de Amparo à Pesquisa do Estado do Rio de Janeiro - Brazil (FAPERJ).

References

Aarthi, N. and K. Murugan. 2012. Effect of *Vetiveria zizanioides* L. Root extracts on the malarial vector, *Anopheles stephensi* Liston. Asian. Pac. Trop. Dis. 2(2): 154–158.

Adams R.P., R.N. Pandey, M.R. Dafforn and S.A. James. 2003. Vetiver DNA-fingerprinted cultivars: effects of environment on growth, oil yields and composition. J. Essential Oil Res. 15(6): 363–371.

Altenor, S., M.C. Ncibi, N. Brehm, E. Emmanuel and S. Gaspard. 2013. Pilot-scale synthesis of activated carbons from vetiver roots and sugar cane bagasse. Waste Biomass Valor. 4: 485–495.

Amarasiri, S.S., A.P. Attanayake, L.D.A.M. Arawwawala, K.A.P.W. Jayatilaka and L.K.B. Mudduwa. 2021. Nephroprotective activity of *Vetiveria zizanioides* (L.) nash supplement in doxorubicin-induced nephrotoxicity model of Wistar rats. J. Food Biochem. 45(9): e13901. doi: 10.1111/jfbc.13901.

Anyango, J.J., D. Bautze, K.K.M. Fiaboe, Z.O. Lagat, A.W. Muriuki, S. Stöckli, et al. 2019. Termite-induced injuries to maize and baby corn under organic and conventional farming systems in the central highlands of Kenya. Insects. 10(10): 367. doi: 10.3390/insects10100367.

Balasankar, D., K. Vanilarasu, P. Selva Preetha, S. Rajeswari, M. Umadevi and D. Bhowmik. 2013. Traditional and medicinal uses of vetiver. J. Med. Plant Studies. 1(3): 191–200.

Barbieri, C. and P. Borsotto. 2018. Essential oils: Market and legislation. pp. 107–127. *In:* Hany A. El-Shemy (ed.). Potential of Essential Oils. IntechOpen. DOI: 10.5772/intechopen.77725.

Belhassen, E., J.J. Filippi, H. Brévard, D. Joulain and N. Baldovini. 2015. Volatile constituents of vetiver: A review. Flavour Fragr J. 30(1): 26–82.

Bruckner, M., T. Hayha, S. Giljum, V. Maus, G. Fischer, S. Tramberend, et al. 2019. Quantifying the global cropland footprint of the European Union's non-food bioeconomy. Environ. Res. Lett. 14: 045011.

Burger, P., A. Landreau, M. Watson, L. Janci, V. Cassisa, M. Kempf, et al. 2017. Vetiver essential oil in cosmetics: What is new? Medicines. 4: 41.

Champagnat, P., G. Figueredo, J-C. Chalchat, A-P. Carnant and J-M. Bessière. 2006. A study on the composition of commercial *Vetiveria zizanioides* oils from different geographical origins. J. Essent. Oil Res. 18: 416–422.

Chaves, T.A. and A.G Andrade. 2013. Capim vetiver (*Vetiveria zizanioides*): produção de mudas e uso no controle da erosão e na recuperação de áreas degradadas. Manual técnico, Programa Rio Rural, Niterói.

Chen, F., X. Wang and H. Kim. 2003. Antioxidant, anticarcinogenic and termiticidal activities of vetiver oil. *In:* Proceedings of the 3rd. International Vetiver Conference. Guangzhou, China, 6–9 October 2003.

Chen, S.W. 1999. Insect on Vetiver hedges. AU J. Technol. 3: 38–41.

Chomchalow, N. and K. Chapman. 2003. Other uses, and utilization of vetiver. Pro. ICV3, Guangzhou, China.

Chou, S.T., C.P. Lai, C.C. Lin and Y. Shih. 2012. Study of the chemical composition, antioxidant activity and anti-inflammatory activity of essential oil from *Vetiveria zizanioides*. Food Chem. 134(1): 262–268.

Chou, S.T., Y. Shih and C.C. Lin. 2016. Vetiver grass (*Vetiveria zizanioides*) oils. pp. 843–848. *In*: V.R. Preedy (ed.). Essential Oils in Food Preservation, Flavor and Safety. Academic Press. London, UK.

D'Amato, D., B. Bartkowski and N. Droste. 2020. Reviewing the interface of bioeconomy and ecosystem service research. AmbioDec.. 12: 1878–1896.

Danh, L.T., P. Truong, R. Mammucari and N. Foster. 2010. Extraction of vetiver essential oil by ethanol-modified supercritical carbon dioxide. Chem. Eng. J. 165(1): 26–34.

Danh, T. 2007. Development of process for purification of α and β-vetivone from Vetiver essential oil and investigation of effects of heavy metals on quality and quantity of extracted Vetiver oil. Ph.D. Thesis University of New South Wales.

David, A., F. Wang, X. Sun, H. Li, P. Lin and G. Deng. 2019. Chemical composition, antioxidant, and antimicrobial activities of *Vetiveria zizanioides* (L.) nash essential oil extracted by carbon dioxide expanded ethanol. Molecules. 24(10): 1897.

Dierz, T., J. Börner, J.J. Förster and J.V Braun. 2018. Governance of the bioeconomy: A global comparative study of national bioeconomy strategies. Sustainability. 10(9): 3190.

Donjadee, S. and T. Tingsanchali. 2012. Reduction of runoff and soil loss over steep slopes by using vetiver hedgerow systems. Paddy Water Environ. 11: 573–581.

Dwivedi, G.R., S. Gupta, S. Roy, K. Kalani, A. Pal, J.P. Thakur, et al. 2013. Tricyclic sesquiterpenes from *Vetiveria zizanoides* (L.) Nash as antimycobacterial agents. Chem. Biol. Drug Des. 82(5): 587–94. doi: 10.1111/cbdd.12188.f

Eapen, S. 2007. Micropropagation of vetiver grass (*Vetiveria zizanoides* L. Nash): A potential plant for environmental conservation. Plant Cell Biotechnol. Mol. Biol. 8(3–4): 193–196.

Emishaw, D. 2017. Extraction, optimization and characterization of essential oil from Vetiver (*Vetiveria zizanioides*) grass root by steam distillation for perfume application. Dissertations, Addis Ababa University. Ethiopia.

Gaspard, S., S. Altenor, E.A. Dawson, P.A. Barnes and A. Ouensanga. 2007. Activated carbon from vetiver roots: Gas and liquid adsorption studies. J. Hazard. Mater. 144(1–2): 73–81.

Grover, M., T. Behl, S. Bungau and L. Aleya. 2021. Potential therapeutic effect of *Chrysopogon zizanioides* (Vetiver) as an anti-inflammatory agent. Environ. Sci. Pollut. Res. Int. 28(13): 15597–15606, doi: 10.1007/s11356-021-12652-z.

Guadarrama, D. 1990. The decline of vetiver production in Reunion. Marches Tropicaux et Mediterraneens, 46: 75–77.

Gupta, C., D. Prakash and S. Gupta. 2015. A biotechnological approach to microbial based perfumes and flavours. J. Microbiol. Exp. 2(1): 11–18.

Gurib-Fakim, A., M. Sewraj, J. Gueho and E. Dulloo. 1993. Medical ethnobotany and some weeds of Mauritius and Rodrigues. J. Ethnopharmacol. 39: 175–85.

Han, X. and T.L. Parker. 2017. Anti-inflammatory activity of clove (*Eugenia caryophyllata*) essential oil in human dermal fibroblasts. Pharm. Biol. 55: 1619–1622.

Hanayama, N., F. Kido, R. Tanaka, H. Uda and A. Yoshikoshi. 1973. Sesquiterpenoids of vetiver oil—I: The structures of zizanoic acid and related constituents. Tetrahedron. 29(7): 945–954. doi.org/10.1016/0040-4020(73)80044-4

Hirunpraditkoon, S. and A.N. García. 2009. Kinetic study of vetiver grass powder filled polypropylene composites. Thermochim. Acta. 482: 30–38.

INC. 2015. Carbon Roots International. Briquettes de Vétiver. Rapport de Faisabilité. Available online: wedocs.unep.org/ bitstream/handle/20.500.11822/22502/Haiti_vetiver_briquette_feasibility.

IndustryarcARC. Mercado de óleo de vetiver—Análise, tamanho, participação, crescimento, tendências e previsão da indústria global 2021–2026. Available online: www.industryarc.com/Report/16665/vetiver-oil-market.html#. Accessed on 16 Jan. 2022.

Ishrath, P.K., A. Kumar and P.M. Nibin. 2019. Agrotechniques for quality planting material production in vetiver (*Chrysopogon zizanioides* (L.) Nash.). J. Med. Plants Stud. 7: 211–213.

Khan, U.G., A. Saeed and M.T. Alam, 1997. Indusyunic Medicine: Traditional Medicine of Herbal, Animal and Mineral Origin in Pakistan. Dept. of Pharmacognosy, Faculty of Pharmacy, Univ. of Karachi, Karachi. Pakistan.

Kim, H.J., F. Chen, X. Wang, H.Y. Chung and Z. Jin. 2005. Evaluation of antioxidant activity of vetiver (*Vetiveria zizanioides* L.) oil and identification of its antioxidant constituents. J. Agric. Food. Chem. 53(20): 7691–5, doi: 10.1021/jf050833e.

Lal, R.K. 2013. On genetic diversity in germplasm of vetiver '*Veteveria zizanioides* (L.) Nash'. Ind. Crops. Prod. 43: 93–98.

Lavania, S. 2003a. Vetiver root system: search for the ideotype. pp. 526–530. *In:* Proceedings of the Third International Conference on Vetiver and Exhibition. Guangzhou. China.

Lavania, U.C. 2003b. Vetiver Root—Oil and its Utilization. PRVN Tech. Bull. No. 2003/1, ORDPB, Bangkok.

Lavania, U.C. 2003c. Other uses of vetiver: Part II. Vetiver oil. *In*: Proceedings of the Third International Conference on Vetiver and Exhibition, Guangzhou, China.

Lawrence, B.M. 2016. Progress in essential oils: 2016. Perfum. Flavorist. Available online: www.perfumerflavorist.com/fragrance/ingredients/article/21856205/progress-in-essential.

Lee, J.H. and J.S. Lee. 2010. Chemical composition and antifungal activity of plant essential oils against *Malassezia furfur*. J. Microbiol. Biotechnol. 38(3): 315–321.

Leite, B. 2012. Extraction of essential oils from vetiver (*Vetiveria zizanioides*) grass. Ph.D. Thesis, University of KwaZulu-Natal.

Lingga, P. 2001. Recipes of Traditional Medicine, 19th Ed. Penebar Swadaya, Jakarta, Indonesia (in Indonesian).

Liu, W., X. Lin, L. Luo, M. Yao, Y. Gu, M. Xie, et al. 2021. Seedling production from seeds of a wild ecotype of Vetiver grass (*Vetiveria zizanioides* L.) in Southern China. Am. J. Plant Sci. 12: 394–405.

Maffei, M. 2002. Vetiveria: the Genus *Vetiveria*. Taylor & Francis.

Mallavarapu, G.R., K.V. Syamasudar, S. Ramesh and B.R.R. Rao. 2012. Constituents of south Indian vetiver oils. Nat. Prod. Commun. 7. 1934578X1200700228.

Martinez, J., P.T. Rosa, C. Menut, A. Leydet, P. Brat, D. Pallet, et al. 2004. Valorization of Brazilian vetiver (*Vetiveria zizanioides* (L.) Nash ex Small) oil. J. Agric. Food Chem. 52: 6578–6584.

Miller, L. 2006. Ayurvedic Remedies for the Whole Family. Motilal Banarsidass, Delhi.

Moussou, P., D. Louis, B. Vincent, J. Christine and P. Gilles. 2016. Cosmetic preparations containing PTH fragments. U.S. Patent # 9,387,235.

Nimityongskul, P., S. Panichnava and T. Hengsadeekul. 2003. Use of vetiver grass ash as cement replacement materials. Proc. icv-3: 148–159. Available online: http://prvn.rdpb.go.th/files/7-06.pdf. Accessed on 10 Aug. 2021.

Oscar, T. 2015. Haïti: Premier Exportateur Mondial de Vétiver. Available online: challengesnews.com/haiti-premier-exportateur-mondial-de-vetiver. Accessed on 08 Aug. 2021.

Oshunsanya, S.O. 2013a. Surface soil properties and maize yields in runoff plots planted with vetiver grass (*Vetiveria nigritana* Stapf) hedges. Soil Science. 178: 205–213.

Oshunsanya, S.O. 2013b. Crop yields as influenced by land preparation methods established within vetiver grass alleys for sustainable agriculture in Southwest Nigeria. Agroecol. Sustain. Food Syst. 37: 578–591.

Oshunsanya, S.O. and E.A. Ewetola. 2011. Production of hygienic *Amaranthus cruentus* on soils remediated with vetiver grass species in Ibadan. Nigeria. J. Sci. Res. 10: 212–220.

Oshunsanya, S.O., K. Are and J.A. Fagbenro. 2014. The use of vetiver grass (*Vetiveria nigritana* Stapf) strips in checking soil loss and improving yields of maize-cassava-cowpea intercropping systems in Southwest Nigeria. J. Appl. Agric. Res. 6: 237–244.

Parisca, O.R. and G.Y. Tamayo. 2006. Vetiver (*Vetiver zizanioides*): una extraordinaria planta para la protección ambiental. Boletín Vetiver. Red Latinoamericana del Vetiver, Maracay–Venezuela.

Parmar, M., P. Shah, V. Thakkar, S. Al-Rejaie and T. Gandhi. 2013. Hepatoprotective potential of methanolic extract of *Vetiveria zizanioides* roots against carbon tetrachlorideinduced acute liver damage in rats. Dig. J. Nanomater. Biost. 8(2): 835–844.

Pereira, A.R. 2006. Uso do vetiver na estabilização de taludes e encostas. Boletim Técnico DEFLOR–Bioengenharia. 1: 1–22.

Pinthong, J., S. Impithuksa and A. Ramlee. 1998. The capability of Vetiver hedgerows on the decontamination of agro chemical residues. Proc. First Inter. Conf. on Vetiver. Thailand. 91–98.

Porras, S. and F. Jochamowitz. 2002. Hair restorer containing vetiver grass extract. U.S. Patent #10/121,795.

Powers, C.N., J.L. Osier, R.L. McFeeters, C.B. Brazell, E.L. Olsen, D.M. Moriarity, et al. 2018. Antifungal and cytotoxicity activities of sixty comercially-available essential oils. Molecules. 23: 1–13.

Pripdeevech, P., S. Wongpornchai and A. Promsiri. 2006. Highly volatile constituents of *Vetiveria zizanioides* roots grown under different cultivation conditions. Molecules. 11: 817–826.

Priyadarshani, N.D., M.K. Amarasinghe, S. Subasinghe, I.R. Palihakkara and H.K. Kumarasingh. 2013. Effect of organic and inorganic fertilizers on biomass production, oil yield and quality of vetiver (*Vetiveria zizanioides* L.). J. Agric. Sci. 8: 28–35.

Raman, J.K. and E. Gnansounou. 2015. LCA of bioethanol and furfural production from vetiver. Bioresour. Technol. 185: 202–210.

Raman, J.K., C.M. Alves and E. Gnansounou. 2018. A review on moringa tree and vetiver grass—Potential biorefinery feedstocks. Bioresour. Technol. 249: 1044–1051.

Rao, R.C., C.S.L. Gal, I. Granger, J. Gleye, J.M. Augereau and C. Bessibes. 1994. Khusimol, a non-peptide ligand for vasopressin V_{1a}, receptors. J. Nat. Prod. 57: 1329–1335.

Rios, J.L. and M.C. Recio. 2005. Medicinal plants and antimicrobial activity. J. Ethnopharmacol. 22(2005): 80–84. doi: 10.1016/j.jep.2005.04.025.

Robbins, S.R.J. 1982. Selected markets for the essential oils of patchouli and vetiver. Report of the Tropical Products Institute. London.

Robert, J.J. 2009. Plant Guide.'Sunshine' Vetivergrass *Chrysopogon zizanioides* (L.). USDA NRCS Plant Materials Center, Hoolehua, Hawaii.

Saikia, D., S. Parveen, V.K. Gupta and S. Luqman. 2012. Anti-tuberculosis activity of Indian grass KHUS (*Vetiveria zizanioides* L. Nash). Complement. Ther. Med. 20: 434–436.

Santos, D.S., J.V. Oberger, R. Niero, T. Wagner, F. Delle Monache, A.B. Cruz, et al. 2014. Seasonal phytochemical study and antimicrobial potential of *Vetiveria zizanioides* roots. Acta Pharm. 64(4): 495–501. doi: 10.2478/acph-2014-0040.

Sarrazin, E. 2017. The scent creation process. pp. 137–138. *In*: A. Buettner (ed.). Springer Handbook of Odor: Springer Handbooks. Springer, Cham. Switzerland.

Sastry, K.N.R. 1998. Socio-economic dimensions of vetiver in rainfed areas of Karnataka, India. Proc. ICV-1, Chiang Rai. Thailand. 243–248.

Shabbir, A., M.M. Khan, B. Ahmad, Y. Sadiq, H. Jaleel and M. Uddin. 2019. *Vetiveria zizanioides* (L.) Nash: a magic bullet to attenuate the prevailing health hazards. pp. 99–120. *In*: M. Ozturk and K.R. Hakeem (eds). Plant and Human Health, 2. Springer, Cham. Switzerland.

Shealy, C.N. 1998. The Illustrated Encyclopedia of Healing Remedies. Element Books Limited. USA.

Silveira, J.C., N. Busato, A. Costa and E. Costa Junior. 2012. Survey and analysis of essential oil extraction methods. Biosphere Encyclopedia. 8: 238–252. Available online: conhecer.org.br/ojs/index.php/biosfera/article/view/3767. Accessed on 10 Oct. 2021.

Skaria, B.P. 2007. Aromatic Plants. New India Publishing.

Smitha, G.R., T.S. Varghese and P. Manivel. 2014. Cultivation of Vetiver [*Vetiveria zizanioides* (Linn)]. ICAR– Directorate of Medicinal and Aromatic Plants Research, Boriavi, Anand.

Sreenath, H., K. Jagadishchandra and Y. Baja. 1994. *Vetiveria zizanioides* (L.) Nash (Vetiver grass). *In vitro* culture, regeneration and the production of essential oils. pp. 403–420. *In*: Y.P.S. Bajaj (Ed.). Biotecnology in Agriculture and Forestry 26. Medicinal and Aromatic Plants VI. Springer-Verlag Berlin Heidelberg GmbH, New Delhi, India.

Suyono, H. and D. Susanti. 2020. The medical benefits of vetiver essential oil. pp. 9–12. Proceedings of the 2nd International Conference of Essential Oils (ICEO 2019). Indonesia.

Talansier, E., M.E.M. Braga, P.T.V. Rosa, D. Paolucci-Jeanjean and M.A.A. Meireles. 2008. Supercritical fluid extraction of vetiver roots: A study of SFE kinetics. J. Supercrit. Fluids. 47: 200–208.

Truong, P. 1993. Report on the international vetiver grass field workshop, Kuala Lumpur. [soil and water conservation; implications for Australia]. Australian J. Soil Water Conserv. 6: 23–26.

Truong, P. 2000. Vetiver grass for mine site rehabilitation and reclamation. pp. 85–86. Proceedings of Remade Lands International Conference. Fremantle. Australia.

Truong, P.N. and D. Baker. 1998. Vetiver grass system for environmental protection. Technical Bulletin. Pacific Rim Vetiver Network. Office of the Royal Development Projects Board, Bangkok. Thailand.

Ubando, A.T., C.B. Felix and C. Wei-Hsin. 2020. Biorefineries in circular bioeconomy: A comprehensive review. Bioresour. Technol. 299: 122585.

Virshette, S.J., M.K Patil and J.R. Shaikh. 2020. A review on pharmacological properties and phytoconstituents of indigenous carminative agents. Int. J. Pharmacogn. Phytochem. 9: 142–145.

Wagner, H. and G. Ulrich-Merzenich. 2009. Synergy research: approaching a new generation of phytopharmaceuticals, Phytomedicine. 16(2–3): 97–110, doi: 10.1016/j.phymed.2008.12.018.

Zhu, B.C., G. Henderson, A.M. Sauer, Y. Yu, W. Crowe and R.A. Laine. 2003. Structure-activity of valencenoid derivatives and their repellence to the Formosan subterranean termite. J. Chem. Ecol. 29: 2695–701.

Section 3

Properties of Essential Oils

9

Toxicity, Anti-nociceptive and Anti-inflammatory Activities of Essential Oils from Medicinal Plants from Nigeria

Opeyemi Nudewhenu Avoseh[1], *Abdulatif Giwa-Ajeniya*[1] *and Isiaka Ajani Ogunwande*[2*]

Introduction

Plant phytochemicals are extremely complex matrices of highly active components with pharmaceuticals, pharmacological and industrial applications. These applications depend on the ethnopharmacological information garnered from historical and scientific reports. In this present study, the pain-mediating potentials of essential oils from some plants grown in Nigeria are described. The use of volatile terpenoids for pain amelioration is due to their high volatility, high penetrating power through skin follicles, increased passage through membranes, and excessive metabolism at pain sites. Reports suggested that essential oils use their lipophilic fractions or sites on lipid moieties of cell membranes, thereby regulate the action of calcium and potassium-ion channels when either inhaled or applied to the skin (Saad et al. 2013, Shah et al. 2011). Analgesic activities of essential oils have been widely reported; they easily penetrate the blood-brain barrier and act in the central nervous system, thereby acting as a psychoactive drug. Further interactions with cell membranes also alter the transfer systems, enzymatic actions or receptors. Carrageenan is used widely to induce paw edema in rodents to validate the anti-inflammatory effect of drugs or herbs. This model is considered as a biphasic model. The early stage (1–2 hr) contributes to the release of histamine, bradykinin, and serotonin which facilitate the increased synthesis of PGs from the surrounding tissues of the injured site. The late phase (3–4 hr) is

[1] Department of Chemistry, Faculty of Science, Lagos State University, Ojo, Nigeria.
[2] University Road, Aleku Area, Osogbo, 230271, Nigeria.
* Corresponding author: isiakaogunwande@gmail.com

characterized by the increased level of PG mediated by the elevated modulation of leukotrienes and bradykinin. During this phase the Cyclo-Oxygenase-2 (COX-2) converts arachidonic acid into PG which is a key factor of the inflammation regulation.

The volatile phytochemicals, essential oils and their components have been known for food preservation, as insect repellents, to alleviate pest infestation, and for ameliorating human illnesses, since ancient times. The essential oil of a plant is the concentrated, volatile, aromatic mixture of chemicals that is formed in various organs or tissues, including leaves, bark, flowers, fruits, and roots. Plants have evolved these volatile chemicals for a number of purposes. It should be noted that several essential oils have exhibited anti-nociceptive and anti-inflammatory activity on a dose-dependent basis and time-dependent basis (Salud et al. 2011). The high structural diversity of the essential oil constituents increases their synergistic effects. Terpenoids are established anti-inflammatory agents which inhibit the production of NOS-2 and COX-2 proteins, and NOS-2, COX-2, TNF-α mRNAs, and NF-κB (Piccinelli et al. 2015). Many essential oils effectively inhibit pro-inflammatory cytokine production. α-Humulene from *C. verbenacea* prevented neutrophil migration and TNF-α production (de las Heras and Hortelano 2009); 1,8-cineole inhibited TNF-α and IL-1β in human lymphocytes (Passos et al. 2007). Terpinen4-ol suppressed the production of TNF-α, IL-1β, IL-8, IL-10, and PGE2 by lipopolysaccharide (LPS)-activated monocytes (Juergens et al. 2004).

In this chapter, the chemical compositions, toxicity, anti-nociceptive and anti-inflammatory potentials of essential oils from some species of plants growing in Nigeria are described.

Collection of the Different Parts of the Plants

Different parts of different plant samples were collected from different locations in Nigeria. The plants are described in Table 9.1. All the plant samples were duly identified and authenticated, while herbarium numbers were deposited at a notable Herbaria within the country.

Isolation of Essential Oils from the Different Parts of the Plants

The different plant samples were processed as air-dried before hydrodistillation. A known weight of each of the was subjected to hydrodistillation in a Clevenger-type apparatus according to established specification (Ogunwande et al. 2019a,b; Avoseh et al. 2020a,b). The essential oils which were distilled in water were collected separately by running through a tap in the receiver arm of the apparatus into clean and previously weighed sample bottles. The oils were kept under refrigeration until the moment of analyses. In most cases, the essential oil yields and color were also given.

Analysis of the Essential Oils

All the essential oils were analyzed by using Gas Chromatographic analysis (GC) and gas chromatograph-Mass Spectrometry (GC/MS) techniques. The

Table 9.1 Plants from Nigeria used in the study of anti-inflammatory and anti-nociceptive activities

Species	Location	Parts	Oil yield/color	References
Petiveria alliacea L.	Ijede, Ikorodu, Lagos state	Leaves	(w/w), colourless	Oluwa et al. 2017
Gardenia jasminoides J. Ellis	Igando, Lagos state	Leaves	0.23% (v/w, ±0.01), colourless	Avoseh et al. 2020a
Albizia zygia (DC.) J.F. Macbr.	University of Lagos, Botanical garden	Leaves	0.1% (w/w), yellow	Avoseh et al. 2021
Albizia lebbeck		Leaves		
Casuarina equisetifolia J.R. & G. Forst	Front of the Okunuga Hall, Faculty of Law, Lagos State University, Ojo	Leaves	0.21% (v/w), colorless	Avoseh et al. 2022
Turnera diffusa Wild	premises of Eletu Odibo Senior Secondary School, Akoka-Yaba, Lagos	Leaves	0.19% (v/w, yellow)	Lawal et al. 2020
		Stem bark	0.28% (v/w, colorless)	
Datura stramonium L.	Odonla village, Ikorodu, Lagos State	Leaves	0.35% (w/w), light yellow	Aboluowodi et al. 2017
		Seeds	0.01% (w/w)	
Jatropha curcas L.	Ijoko, Ogun State	Leaves	0.60%, w/w, yellow	Adeosun et al. 2017
Cordia millenii (Bak.)	Ayetoro, Ilesha, Osun State	Root barks	0.11 (v/w, colourless)	Avoseh et al. 2019a
Mucuna pruriens Utilis	Imota Ikorodu Forest Reserve, Lagos State	Leaves	0.2 (v/w), light-yellow	Avoseh et al. 2020b
Bougainvillea glabra Choisy	Staff Garden, University of Lagos	Leaves	0.08% (v/w)	Ogunwande et al. 2019a
Phyllanthus muellerianus (O. Kutze) Exell	Onigambari Forest Reserve, Oyo State	Leaves	0.90% (v/w), pale yellow	Ogunwande et al. 2019b
Waltheria indica L.	Botanical Garden, Lagos State University	Leaves	0.08% (v/w), light green	Avoseh et al. 2019b
Moringa oleifera Lam	University of Ilorin Plantation	Flowers, leaves, seeds, barks, and roots	0.3%–0.36%	Otunola and Afolayan, 2018
Dennettia tripetala G. Baker	Owena market, Owena Town, Ondo State	Fruits	3.7% w/w	Oyemitan et al. 2008

identification of constituents of essential oils from the GC/MS spectral was performed by the comparison of Retention Indices (RI Exp.) with reference to a homologous series of *n*-alkanes, under identical experimental conditions. In some cases, co-injection with known compounds under the same GC conditions was used. The mass spectral (MS) fragmentation patterns were checked with those of other essential oils of known composition in literature (Ogunwande et al. 2019a,b, Avoseh et al. 2020a,b).

Antinociceptive and Anti-inflammatory Tests

The hot plate method was used to determine the anti-nociceptive property, whereas the anti-inflammatory activity was evaluated by an egg albumin (Adeosun et al. 2017), carrageenan-induced and formalin experimental models as described in various studies (Ogunwande et al. 2019a,b, Avoseh et al. 2020a,b), as well as denaturation of BSA and rat erythrocyte membrane hemolysis (Otunola and Afolayan 2018).

Results and Discussion

Jatropha curcas L.

The chemical constituents and anti-inflammatory activity of the essential oil from the leaf of *Jatropha curcas* L (Euphorbiaceae) collected from Nigeria was evaluated recently. The analysis of the chemical constituents revealed that all the constituents of the essential oil, namely neophytadiene (35.8%), phytol (23.1%), trans-pinane (12.7%), 6,10,14-trimethyl-2-pentadecanone (12.3%) and citronellyl propanoate (11.2%), were present in significant amounts. The incremental percentage inhibition of inflammation by the leaf essential oil of *J. curcas* was investigated on Wistar rats using egg albumin as a phlogistic agent. The observed inhibition progression namely 20.0% (1st hr), 36.8% (2nd hr), 50.0% (3rd hr) and 76.6% (4th hr) was as a result of different levels of absorption of phytochemicals responsible for the anti-inflammatory effect in *J. curcas* oil. The anti-inflammatory activity of the essential oil reveals a minimal to significant activity. The inflammation was studied using an egg albumin as a phlogistic agent due to its proteinuous nature. As expected, the standard drug (ibuprofen) exhibited inhibition ($p < 0.001$) compared with the control (water). The essential oil of *J. curcas* gave a significant inhibition of $p < 0.05$ consistent throughout the 1st to the 4th hr of experimentation (Adeosun et al. 2017). The significant activity of the oil may be due to the presence of phytochemicals such as phytol (Avoseh et al. 2015, Silva et al. 2014) and neophytadeiene (Carretero et al. 2008), which are known for their anti-inflammatory potential.

Petiveria alliacea L.

The air-dried leaves of *Petiveria alliacea,* collected from Ijede, Ikorodu, Lagos State, Nigeria were hydrodistilled and the essential oil screened for the anti-inflammatory property. Eleven compounds representing 99.1% of the oil content were identified in the essential oil of *P. alliacea*. The main constituents were phytol (56.1%), citronellol (16.0%) and (*Z, Z*)-α-farnesol (14.6%). In the egg albumin-induced edema, essential oils of *P. alliacea* induced a time dependence in paw edema. The anti-inflammatory activity was non-significant ($p = .05$) from the 1st hr to the 3rd hr after the egg albumin injection at 1 mL of 2% dose level while the maximum inhibition ($p < .001$) was observed after 4 hr. At this hour, the inhibition was more pronounced than Diclofenac sodium injection (Oluwa et al. 2017).

The present activity may be due to the constituents of the essential oil such as phytol which significantly reduced carrageenan-induced paw edema, in a dose dependent manner, inhibited histamine, serotonin, bradykinin and PGE2 induced paw edema (Silva et al. 2014). Earlier results suggest farnesol supplementation may be beneficial to improve the skewed allergic asthmatic inflammation (Ku and Lin 2015), while the antinociceptive and anti-inflammatory properties of citronellol in rodents have been documented (Brito et al. 2012).

Gardenia jasminoides J. Ellis

The chemical constituents, anti-inflammatory and anti-nociceptive activities of the leaf essential oil hydrodistilled from the *Gardenia jasminoides* J. Ellis (Rubiaceae) were documented earlier. The chemical constituents of the essential oil were determined by using gas chromatography-Flame Ionization Detector (GC-FID) and gas chromatography coupled with mass spectrometry (GC/MS) revealed the abundance of pentadecanal (49.2%), geranial (12.3%), ar-turmerone (8.2%) and 10-epi-γ-eudesmol (6.2%). The hot plate and carrageenan-induced models were used to determine the anti-nociceptive and anti-inflammatory activities. The acute toxicity of the essential oil which was evaluated at 500, 1000, 1500, and 2000 mg kg body weight showed no contrary effects on the behavioral responses in the tested rats following 14 d of observation. No mortality, size, or weight change observed. Therefore, the highest dose of 400 mg/kg adopted to be ministered to rats in this study was considered safe (Avoseh et al. 2020a).

The leaf essential oil of *G. jasminoides* at 100 and 400 mg/kg doses significantly ($p < 0.001$) increased the latency period for the reaction duration in the anti-nociceptive study. The carrageenan-induced edema model reveals the suppression of inflammatory mediators ($p < 0.001$) within 1st, 2nd and 4th hr for 100 mg/kg dose, 1st and 4th hr (200 mg/kg) and 1st–3rd hr (400 mg/kg). The results indicate the potential of *G. jasminoides* essential oil as a source of pain relieving agent (Avoseh et al. 2020a). For the anti-nocieptive activity, it was observed that the higher the maximum pain threshold time, the more the inhibitory capacity of *G. jasminoides* essential oil. At the 100 and 400 mg/kg, doses of *G. jasminoides* significantly ($p < 0.001$) increased the latency period for the reaction duration. Conversely, the 200 mg/kg did not show any significant inhibition (Avoseh et al. 2020).

The abundant constituents of the essential oil mainly pentadecanal, geranial, *ar*-trumerone, etc. were earlier known for their pharmacological properties. Geranial exhibits significant anti-inflammatory and analgesic activities, inhibiting expression of the inflammatory mediators such as TNF-α and IL-6 secretion of macrophages, pro-IL-1β, iNOS, COX-2 and NLRP-3, phosphorylation of ERK1/2, JNK1/3, p38 and IκB e.t.c., (Mai et al. 2014, Nishijima et al. 2014, Liao et al. 2015). *ar*-Turmerone also reduced TNF-α, IL-1β, IL-6, and MCP-1 production in amyloid (Aβ) stimulated microglial cells as well as NF-κB, JNK, and p38 MAPK signaling pathways (Park et al. 2012a,b, Yang et al. 2020). However, it is difficult to explain the possible mechanism of the essential oils based on the present findings; further studies are needed to elucidate the exact mechanism.

Albizia zygia and Albizia lebbeck

The essential oils from the leaves of *Albizia zygia and A. lebbeck* were obtained in a yield of 0.1% (v/w) calculated on a dry weight basis for both oils. The main compounds of *A. lebbeck* were 2-pentylfuran (16.4%), (*E*)-geranyl acetone (15.46%), (*E*)-α-ionone (15.45%) and 3-octanone (11.61%), while the oil of *A. zygia* is mainly hexahydrofarnesyl acetone (33.14%), (*E*)-methyl isoeugenol (11.7%) and 2-methyl tetradecane (6.64%). The volatile oils are non-toxic for about 5000 mg/ kg dose. The heat latency capacity of the essential oils was analyzed using the hot plate model. The essential oil *A. lebbeck* was not effective in blocking the nociceptive response to thermal stimuli compared to the saline-treated group for all the doses evaluated at all reaction times. However, a high rate of inhibition was displayed by the *A. zygia* essential oil. The 100 mg/kg dose displayed an increased activity ($p < 0.01$–0.001) as the reaction time increased. Increasing the dose to 200 mg/kg, significant inhibition ($p < 0.001$) of the thermal stimuli was reported for the 30th and 60th min, which decreased slightly to a minimal ($p < 0.01$) at the 90th min. The increased latency capacity of essential oils was synonymous with the high value of the maximum pain threshold. The 400 mg/kg dose of *A. zygia* shows a slight decrease in activity as reaction time increases. The 30th min was highly significant ($p < 0.001$), but a steady decline to $p < 0.01$ at 60th and 90th min. This is an indication that *A. zygia* possesses a very high anti-nociceptive potential than *A. lebbeck* (Avoseh et al. 2021).

Both essential oils displayed a relatively moderate inhibition against the Formalin at the neurogenic phase (0–5 min). In terms of the dose, the 100 mg/kg of the *A. zygia* shows the highest inhibitory capacity (100%), displaying an activity more elevated than the standard drug. In contrast, the same dose of *A. lebbeck* reveals a minimal activity of 3%. *A. lebbeck* oil's inhibitory capacity at the neurogenic phases was an average of 47% for the 200 and 400 mg/kg doses. In phase II (anti-inflammatory phase, 15–30 min), both oils showed maximal activities and acted like the piroxicam drug. The inhibitory activities are independent of the dose and the type of *Albizia* drug evaluated in this study. This study shows that *A. zygi* significantly antagonized the proliferation of thermally mediated receptors more than *A. lebbeck* (Avoseh et al. 2021).

For the anti-inflammatory test, orally, administering *A. lebbeck* and *A. zygia* 1 hr before a carrageenan injection exerted a significant dose-dependent inhibitory effect on the development of carrageenan-induced paw swelling. Furthermore, the most active doses observed in *A. lebbeck* were the 100 and 200 mg/kg, whose activities remain constant from the 1st to the 4th hr post-carrageenan at a significant rate of $p < 0.001$. The 400 mg/kg displayed a decreased pro-inflammatory mediator inhibition from the 1st to the 4th hr ($p > 0.5 - p < 0.01$). *A. zygia* exerted a highly significant ($p < 0.001$) non-dose dependent inhibitory effect at 1–3 hr. However, at the 4th hr, a decreased activity was reported for the 100 mg/kg ($p > 0.05$) and 400 mg/kg ($p < 0.05$). The 200 mg/kg *A. zygia* shows the highest activity during the anti-inflammatory assay (Avoseh et al. 2021).

The marked immunological changes observed in the current study can be attributed to the plant species' chemical constituents whose activities could be in synergy or separated. The presence of apocarotenoids such as *E*-(α)-ionone, *E*-geranyl acetone (*A. lebbeck*), hexahydrofarnesyl acetone, and (*E*)-methyl isoeugenol (*A. zygia*) are products of the cleavage of parent carotenoids exclusively produced by plants and microorganisms. They possess conjugated double bond systems conferring on them efficient antioxidant, anti-inflammatory and anti-cancer properties (Harrison and Quadro 2018). α-Ionone has been shown to increase procollagen content of human dermal fibroblasts and could attenuate UVB-induced loss of collagen and Hyaluronic Acid (HA), thereby improving edema healing and ageing (Belsito et al. 2007). A β-ionone significantly inhibits the expression of LPS-induced pro-inflammatory mediators such as NO, PGE2 and TNF-α, as well as their regulatory genes such as iNOS, COX-2 and TNF-α in BV2 by controlling of Akt-dependent NF-jB activity and the MAPK pathway (Kang et al. 2013). Compounds containing the α-β-unsaturated carbonyl moiety inhibit pro-inflammatory mediators by downregulating the production of TNF-α, COX-2, NOs2 and PGE-2 through the Michael addition reactions in biological systems (Kang et al. 2013). Eugenol and its isomers are one of the most outstanding phenylpropanoids with high anti-nociceptive and anti-inflammatory properties. Tang et al. (2015) demonstrated that methyl eugenol significantly suppresses degranulation and pro-inflammatory cytokine release in antigen-sensitized mast cells . Inhibition of Ca^{2+}, Na^{+}, and K^{+} currents are mechanistic pathways in which methyl eugenol displays anti-nociceptive activities. Methyl eugenol inhibits peripheral nerve Nav1.7 channels with a higher affinity for inactivated and open channels and a weaker affinity to those in the resting state (Wang et al. 2015). The specific activities displayed by the monoterpenoids and other components further provide a piece of evidence for the actions shown by ALEO and AZEO.

Casuarina equisetifolia J.R. & G. Forst

The pale-yellow essential oil of *Casuarina equisetifolia* was obtained in yield of 0.21% (v/w), calculated on a dry weight basis. The main constituents of the essential oil were methyl salicylate (30.4%), a-zingiberene (15.5%), (*E*)-anethole (9.5%), b-bisabolene (8.6%), b-sesquiphellandrene (6.9%), and *ar*-curcumene (6.2%). In this study, the results obtained at test doses of 500, 1000, 1500 and 2000 mg/kg body weight demonstrated a high safety margin with no sign of divergent effects on the reflex, health, weight, food and water intake habit and no mortality recorded for 14 d. Therefore, the highest dose of 400 mg/kg given to rats in this study was regarded as safe (Avoseh et al. 2022). The reaction of Wistar rats to heat was recorded from the time zero after administration of the essential oil to the 90th min, to understudy the effect on heat resistance by the rats. In this result, 200 mg/kg extract of the essential oil significantly increased the thermal resistance of the rats throughout the reaction period at a significant value of $p < 0.001$ (0th to 60th min) and $p < 0.01$ for the 90th min. The lowest activity was recorded by the 100 and 400 mg/kg of essential oil, where inhibition was only significant at the 60th min and 30th min, respectively ($p < 0.01$). Therefore, the anti-nociceptive property of the essential oil of *C. equisetifolia* was more pronounced in the 200 mg/kg dose independently of reaction time (Avoseh et al. 2022).

To analyze the effect of *C. equisetifolia* essential oil on both the peripheral and centrally mediated analgesia, the formalin-induced assay was adopted. The results depicted the rate of inhibition at the analgesic (anti-nociceptive) and anti-inflammatory stages after the administrations of the essential oil. Elevated inhibitions were observed at the 100 and 200 mg/kg dose of the essential oil, while the 400 mg/kg was relatively lower at the 1st phase. The rate of inhibition of the formalin by the standard was lower (46.81%) when compared to the essential oil at doses of 100 mg/kg (60.28%) and 200 mg/kg (63.83%). At the anti-inflammatory phase, the inhibition rate varies as the concentration changes. Essential oil of *C. equisetifolia* at a dose of 200 mg/kg showed a minimal inhibition of 33% more than other tested doses, like the neurogenic phase. The essential oil also exhibited similar inhibitory activities as the standard drug for the late (anti-inflammatory) phase (Avoseh et al. 2022).

The Carrageenan-induced paw edema was adopted in the study of anti-inflammatory activity. This is a well-defined model for acute inflammation and has been applied in the study of the anti-edematous effect of extracts due to the production of different inflammatory mediators in Wistar rats. The Carrageenan-induced model is a biphasic model leading to the release of inflammation mediators at different hours. The essential oil of *C. equisetifolia* displayed very high inhibitory activities between the 1st to the 3rd hr at a statistical significance value of $p < 0.001$ for all doses evaluated. However, the inhibition rates decrease exponentially as the experiment approaches the 4th hr post-administration. The standard drug (Ibuprofen), 100 and 400 mg/kg were all non-significant ($p > 0.05$) at this hour, however, the 200 mg/kg of the essential oil slightly displayed an inhibition rate of $p < 0.05$ (Avoseh et al. 2022).

The anti-nociceptive and anti-inflammatory inhibitory capacities of *C. equisetifolia* observed in this study can be predicated on the terpenoids present in the essential oil. The major constituents identified in the essential oil of *C. equisetifolia* were known for their biological potentials. The action of methyl salicylate in ameliorating pain has been well documented. Methyl salicylate glycoside inhibited the production of inflammatory cytokines, blocks the activation of the NF-κB signaling pathway and non-selectively suppressed the activity of COX (cyclooxygenase) (Xin et al. 2014). Methyl salicylate was also reported to inhibit the release of lipopolysaccharide (LPS)-induced interleukin (IL)-6 and Tumor Necrosis Factor (TNF)-α in a dose-dependent manner on evaluation of the anti-inflammatory activity derivatives (Li et al. 2016). Essential oil with large contents of α-zingiberene, *ar*-curcumene and sesquiphellandrene produced a significant reduction in acute inflammation produced by carrageenan, dextran and formalin-induced chronic inflammation, as well anti-nociceptive property with a significant reduction in acetic acid-induced writhing movements

(Jeena et al. 2013). Zingiberene exhibits anti-inflammatory property with a significant reduction in the level of inflammatory markers (TNF-α, IL-6, NF-κB, and IL-1β) in rats (Li et al. 2020). Anethole was reported to show moderate anti-inflammatory activity (Freire et al. 2006), and displayed anti-nociceptive effect in the writhing model induced by acetic acid, in the second phase of the formalin test (Ritter et al. 2014). The inhibition of the pro-inflammatory mediators and nociceptive activities by the essential oil is suggestive of the synergistic action of the compounds present in the essential oil.

Turnera diffusa Wild

The essential oils from the leaves and stem bark of *Turnera difusa* were separately isolated using hydrodistillation of the pulverized materials in an all-glass Clevenger-type apparatus and characterized by GC-FID and GC-MS. The hot plate method was used to determine the antinociceptive property, while the anti-inflammatory activity was established using the carrageenan induced paw-edema model. The yields of the essential oils were 0.19% (v/w, yellow, leaf) and 0.28% (v/w, colorless, stem bark), calculated on a dry weight basis. The leaf oil consists mainly of β-caryophyllene (43.7%) and germacrene B (21.3%), while geranial (50.7%) and neral (33.6%) were the main constituents of the stem bark oil.

In the anti-inflammatory study, the essential oil from the stem bark displayed high anti-inflammatory activity ($p < 0.001$) at the 1st to the 4th hr for all the essential oil tested doses. Essential oils from the leaves of *T. diffusa* inhibited inflammation up to the 3rd hr post-treatment, but were non-significant at the 4th hr for all doses (Lawal et al. 2020). The reduction in activity at longer hours (4th hr) for the leaf oil could be a result of the high absorption capacity of the oil by the macrophages because of its high volatility and its inhibitory capacities. For the anti-nociceptive activity, the essential oils of *T. diffusa* (100, 200, and 400 mg/kg) exhibited an anti-nociceptive effect at all observation periods at a significance of $p < 0.001$ for both the leaf and stem bark. The inhibition capacity only reduced at the 120th min for the leaf ($p < 0.01$). The observed analgesic activity in this study likely suggests the application of *T. diffusa* for the treatment of pain-related ailment (Lawal et al. 2020).

In this study, β-caryophyllene, germacrene B, geranial, and neral were found to be the most abundant constituents of the leaf and stem bark essential oils of *T. diffusa.* These terpenoids could be responsible for the observed anti-inflammatory and anti-nociceptive activities of the essential oils. β-Caryophyllene has been shown to suppress inflammation by inhibiting the synthesis of cytokines, reduction in neutrophil, and of IL-6 (Cho et al. 2007, Nishijima et al. 2014, Vijayalaxmi et al. 2015). The mixture of stereoisomers geranial (*E*-isomer) and neral (*Z*-isomer), exhibit significant anti-inflammatory and analgesic activities, inhibiting expression of the inflammatory mediators such as TNF-α and IL-6 secretion of macrophages, pro-IL-1β, iNOS, COX-2 and NLRP-3, phosphorylation of ERK1/2, JNK1/3, p38, and IêB (Mai et al. 2014, Nishijima et al. 2014). However, it is difficult to explain the possible mechanism of the essential oils based on the present findings; further studies are needed to elucidate the exact mechanism.

Datura stramonium L.

Phytol (72.5%) was identified as the main constituent of the leaf essential oil of *Datura stramonium*. However, citral (26.5%), 4,8-dimethyl-3,8-dien-2-one (11.2%), sesquirosefuran (11.1%) and geraniol (10.5%) were identified as the major constituents in the seed oil (Aboluowodi et al. 2017). The anti-inflammation effects of essential oils of *D. stramonium* were tested on fresh egg albumin-induced inflammation measured at 1, 2, 3, and 4 hr. Within the hours, the essential oil activity showed significant ($p < 0.01$) anti-inflammatory as compared to the control. The percentage inhibition shows 22.2, 29.7, 53.8 and 92.3% for 1st to 4th hr respectively (Aboluowodi et al. 2017).

Essential oils from the leaves inhibited inflammation beyond the 4 hr post treatment, while the seed oil was inactive. The potency of the leaf oil shows it has a very high potency to increase the mediation of histamines, serotonin, prostaglandins, and bradykinins due to its high activity over the three (3) phases of acute inflammation. The biological activity of an essential oil has been attributed to the potency of the major compounder a synergy between both the major and minor compounds. The presence of large amounts of phytol in the essential oil of the leaf might be the factor responsible for the anti-inflammatory activity. Diterpene has a strong anti-inflammatory effect by inhibiting the COX1 and the leukotriene (Avoseh et al. 2015, Silva et al. 2014). In addition, leaf ketones also exhibit this property which in this case could be attributed to the 6,10, 14-trimethy-2-pentadecanone and (*E*)-β-ionone present in appreciable amounts in the oil. The potent anti-inflammatory activity of essential oils of *D. stramonium* hereby confirmed its traditional use in treating various inflammatory diseases and may serve as an exploitation of *D. stramonium* in treating various inflammatory diseases.

Cordia millenii (Bak.)

At the test doses of 500, 1000, 1500 and 2000 mg/kg body weight, *Cordia millenii* root bark essential oil, showed no adverse effects on the behavioral and physical responses in the tested rats following an observation for 14 d. There was no mortality, flesh or skin peeling, swollen limbs or neck, and no weight loss was observed. Therefore, a higher dose of 400 mg/kg given to rats in this study was regarded to be safe. The evaluation of the anti-inflammatory activity *in vivo* was conducted using the model of carrageenan-induced paw edema. Edema formation in the paw is the result of a synergism between various inflammatory mediators that increase the vascular permeability and/or the mediators that increase blood flow (Avoseh et al. 2015). This is a well-defined model of acute inflammation 141 and has been applied in the study of the anti-edematous effect of extracts due to the production of different inflammatory mediator in the Wistar rat. This development is time dependent and is characterized by biphasic release of mediators.

The anti-inflammatory activity of *C. millenii* root essential oil was statistically ($p < .001$) for the 100 and 200 mg/kg p.o., at 1st and 3rd hr, while the 200 mg/kg p.o., also displayed significant activity ($p < .001$) at 2nd hr. The 400 mg/kg p.o. displayed anti-inflammatory actions ($p < .001$) at 3rd and 4th hr. The anti-inflammatory activity was also statistically ($p < .01$) for the 100 mg/kg p.o at 2nd hr, while the activity was significant ($p < .05$) for 100 and 200 mg p.o. (4th hr) and 400 mg/kg p.o (1st hr). The anti-inflammatory inhibitory activity of *C. millenii* root oil was greatly significant. The mediators released in all phases were significantly inhibited. The oil activity at these doses was also equivalent to that of the standard drug used (Ibuprofen). However, at the 4th hr, there was significant reduction in the inhibitory activities of the 100 and 200 mg/kg doses. The carrageenan-induced paw edema in rats is believed to be biphasic. In addition, the essential oil of *C. millenii* showed a very prominent activity at all doses. The ability of the essential oil to inhibit the expressions of the nociceptive neurons was very significant at all doses ($p < 0.001$). The activity of the oil showed a similar activity as the standard drug (Acetyl Salicylate), due to statistical significance (Avoseh et al. 2019a).

Mucuna pruriens Utilis

The volatile constituents, toxicity, antinociception, and anti-inflammatory activities of the essential oil obtained from the leaf of *Mucuna pruriens* collected from Nigeria was studied and reported. The anti-nociceptive and anti-inflammatory assays were analyzed by a hot plate, formalin, and carrageenan-induced edema assays, respectively. The essential oil was obtained in a yield of 0.2% (v/w) calculated on a dry weight basis. The essential oil contained a high content of (*E*)-2-hexenal (19.0%), linalool (8.9%), 1-hexanol (6.6%), and *trans*-hydroxylinalool oxide (5.2%).

The acute toxicity of the essential oils evaluated at 500, 1000, 1500, and 2000 mg/kg body weight showed no contrary effects on the behavioral responses in the tested rats following 14 d of observation. No mortality, size, or weight change observed. Therefore, the highest dose of 400 mg/kg was adopted to be administered to rats in this study were considered safe (Avoseh et al. 2020b). The analgesic property of the essential oil was slightly significant ($p < 0.5$) only at the third hr for the 400 mg/kg, while the other doses are less active. The reactions of *M. pruriens* induced Wistar rats to formalin inhibition were tested. At the early phase (neurogenic), the essential oils show a low to moderate inhibition averaged between 31.9 and 54.6%, which is relatively small. The highest peak is by the 400 mg/kg essential oils dose with an inhibition rate of 54.6%. The reduced activities recorded are in agreement with the hot-plate model. At the latter phase (anti-inflammatory), the essential oils significantly inhibited the release of inflammation mediators at a maximal value for all the doses >95.6%. Since the essential oil of *M. pruriens* was effective in suppression of the latter phase of the formalin test, it was seen that the anti-inflammatory activity of *M. pruriens* is mediated peripherally.

In the anti-inflammatory experiment, *M. pruriens* essential oil significantly inhibits (100%) the development of edema at all stages of ($p < 0.001$) exposure except for the 400 mg/kg dose. The results show that *M. pruriens* essential oil exhibit the maximum activity within the 1–3 hr of administration, indicating its ability to suppress the expression of histamine, serotonin, and bradykinin (0–1 hr post-treatment), while the latter phase is suspected to be the inhibition of leukocytes and elevated prostaglandins biosynthesis (second to the fourth hr post-treatment). The reduction in activity at longer hours can be a result of the high absorption capacity of the oil by macrophages because of its high volatility and its inhibitory capabilities. The potential proinflammatory mechanism might be due to effects on several proinflammatory mediators, including, histamine, serotonin, and bradykinin, and the ability of the essential oils to act as centrally mediated opioid analgesic. *Mucuna pruriens* essential oils displayed a high anti-inflammation potential and can be used as a potential centrally mediated opioid antagonist against analgesia (Avoseh et al. 2020b).

The activity of essential oils against inflammatory conditions is mainly attributed to the plant's major constituents or in synergy with other constituents. The volatile oils of *Mimusops elengi* flower containing *E*-2-hexenal as one of the major products selectively inhibited the cyclooxygenase; COX-1 and COX-2 at 66 and 27% inhibition rate, respectively (Amir et al. 2012).

Bougainvillea glabra Choisy

The main constituents of the essential oil hydrodistilled from the leaves of *Bougainvillea glabra* were (*E*)-nerolidol (31.4%), (*E*)-β-ionone (10.3%) and linalool (10.1%). The essential oils in test doses of 500, 1000, 1500 and 2000 mg/kg body weight showed no contrary effects on the behavioral responses in the tested rats following 14 d observation. No mortality or weight change observed. Therefore, a highest dose of 400 mg/kg given to rats in this study was regarded to be safe (Ogunwande et al. 2019a). The anti-nociceptive property of the essential oil was statistically significant $p < 0.001$ at all doses when compared to the control while exhibiting an activity in tandem with the standard drug. The results showed that there was no significant difference on the thermal stimulus in rats treated with normal saline (negative control) throughout the 120 min observation. In comparison to the saline treated animals, a significant increase in the reaction time to thermal pain was detectable in both sodium salicylate and essential oil throughout 120th min. However, the observation in essential oil treated animals increased for the experimental duration. As observed, the minimum pain threshold increased proportionally with the concentration and reaction time. All the tested concentrations of essential oil were highly significant with a maximal value of $p < 0.001$ denoting a highly active analgesia extract.

Treatment with *B. glabra* essential oil reduced the paw edema in a dose-dependent manner at 1, 2, 3, and 4 hr after an injection of carrageenan. The significant activities of the dosed essential oils were inconsistent. The 100 mg/kg *B. glabra* essential oil displayed a steady activity at the 1st and 2nd hr ($p < 0.001$), but reduced considerably at the 3rd hr ($p < 0.05$) and a total loss of activity at the 4th hr ($p > 0.05$). The carrageenan-induced edema reveals a similar activity for the 200 mg/kg *B. glabra* essential oil but with a better activity at the 3rd hr ($p < 0.01$). The result showed that the 400 mg/kg *B. glabra* essential oil was not significant and showed less inhibitory activities as expected throughout the reaction duration. As expected, the reference drug, Diclofenac (5 mg/kg), caused a significant inhibition of post carrageenan edema. The results of this study showed that *B. glabra* essential oil suppressed the rat hind paw edema induced by intra-plantar injection of carrageenan, signifying marked anti-acute inflammatory efficiency. The present study showed that *B. glabra* essential oil produced anti-inflammatory effects in carrageenan-induced rat paw edema in a dose-dependent manner. There are significant inhibition of histamine and serotonin at the 1st by all the dose concentration used except for the 400 mg/kg. Similarly, prostaglandins and other cytokines were also inhibited by the 100 and 200 mg/kg showing a steady activity. The results confirmed earlier findings that *B. glabra* essential oil exhibits a noticeable anti-inflammatory effect in accordance with traditional uses of the plant (Ogunwande et al. 2019a).

Phyllanthus muellerianus (O. Ktze) Exell

The leaves of *Phyllanthus muellerianus* was hydrodistilled and the essential oil characterized by gas chromatography-flame ionization detection and gas chromatography coupled with mass spectrometry analyses. The anti-inflammatory activity was evaluated on carrageenan-induced rat paw edema, while the anti-nociceptive test was based by the hot plate analysis. The hydrodistillation provided 0.09% (dry weight basis) of pale yellow oil. Thirty-eight compounds representing 94.8% were identified in the oil. The major components of the oil were hexahydrofarnesyl acetone (11.6%), isocaryophyllene (9.8%), and limonene (9.4%). The essential oils in test doses of 500, 1000, 1500, and 2000 mg/kg body weight showed no contrary effects on the behavioral responses in the tested rats following 14 d observation. No mortality or weight change was observed. Therefore, a highest dose of 400 mg/kg given to rats in this study was regarded to be safe (Ogunwande et al. 2019b). The results of the anti-nociceptive effect of *P. muellerianus* oil using the hot plate method are presented showing the maximum percentage analgesia of the oil samples in relation to dose and time. The analysis was conducted based on dose dependent of 100, 200, and 400 mg/kg of the body weight of the rat. Statistical significance of $p < 0.05$ was used to test the level of confidence of the values. All test doses increased analgesia and pain resistant of Wistar rats. Compared to the lower dose (100 mg/kg), the 200 and 400 mg/kg dose of *P. muellerianus* oil showed analgesic effect against the hot plate response in rats ($p < 0.001$). Aspirin (100 mg/kg) also showed a potent analgesic effect as expected. In the study, the activities of the 200 and 400 mg/kg *P. muellerianus* oil were comparable and even exceeded the analgesic effect of aspirin at the 30th to the 90th min. The activity of the 100 mg/kg oil increased steadily within the 60th to the 120th min ($p < 0.01$). The percentage analgesia showed a steady increase in the activity of the standard drug (aspirin) which increased relatively to reaction time (17–60%).

The 200 and 400 mg/kg *P. muellerianus* oil exhibited very high activity at the 1st and 2nd hr within the range of 57 to 95%. However, the activity reduced at the 3rd and 4th hr (Ogunwande et al. 2019b).

In the carrageenan-induced acute inflammation, *P. muellerianus* oil-treated groups significantly reduced the mean maximal swelling attained at 1 to 4 hr ($p < 0.001$) for all the doses. As expected, the reference drug, Diclofenac (100 mg/kg), caused a significant inhibition of post carrageenan edema. The study found that the *P. muellerianus* oil (100, 200, and 400 mg/kg) exhibited anti-nociceptive effect in the hot plate-induced pain. In addition to the activity, the effect was prolonged showing a high rate of absorption and inhibition of the pain receptors. These observed effects could be due to the inhibition of nociceptors. *Phyllanthusmuellerianus* oil significantly inhibited the edema formation; 1, 2, 3, and 4 hr after the injection of carrageenan in the hind paw of rats. These results indicated that the phytoconstituents present in *P. muellerianus* oil most likely inhibited the inflammatory mediators of the initial as well as the late phase of inflammation induced with carrageenan (Ogunwande et al. 2019b). In the present investigation, *P. muellerianus* oil revealed the presence of hexahydrofarnesyl acetone and limonene. Limonene, a monoterpenoid possesses strong anti-inflammatory and analgesic properties by preventing the production of NO, PGE-2, IL-1β, IL-6, and TNF-α, as well as the expression of iNOS and COX-2 proteins in LPS-treated RAW 264.7 cells, suggesting that it down-regulated pro-inflammatory mediators and cytokines (Hart et al. 2000). The relative anti-inflammatory observed in essential oil could be attributed to the inhibition of the pro-inflammatory mediators in a synergistic effect of all the constituents present in the essential oils.

Waltheria indica L.

The leaf essential oil of *Waltherica indica* was hydrodistilled and characterized by gas chromatography-Flame Ionization Detection (GC-FID) and gas chromatography coupled with mass spectrometry (GC-MS) analyses. The anti-inflammatory activity was evaluated on carrageenan induced rat-paw edema, while the anti-nociceptive test was based on the hot plate model. The hydrodistillation provided 0.41% (dry weight basis) of light green oil. Forty compounds representing 99.8% were identified in the oil. The main constituents of the oil were limonene (34.7%), sabinene (21.2%) and citronellal (9.7%). Test doses of 500, 1000, 1500 and 2000 mg/kg body weight of *W. indica* essential oil showed no adverse effects on the behavioral and physical responses in the tested rats following an observation for 14 d. There was no mortality, flesh or skin peeling, swollen limb or neck, and no weight loss was observed. Therefore, a higher dose of 400 mg/kg given to rats was regarded to be safe (Avoseh et al. 2019b). The essential oils showed high activity for all the doses when compared with the control (saline solution). The 100 mg/kg *W. indica* essential oil displayed a time dependent activity at the 30th min ($p < 0.01$) and ($p < 0.001$) at the 60th min, but inhibition was reduced considerably to a non-significant (ns) value at the 90th and 120th min. The higher doses (200 and 400 mg/kg) displayed high inhibitory activity of thermal stimuli and were of statistical significant of $p < 0.001$ during the entire period of exposure to heat sensitivity. The results indicate that *W. indica* essential oil had a significant ability to prolong the response latencies indicating a significant increase of the nociceptive threshold in treated animals as compared to the untreated groups typical of opioid drugs (Avoseh et al. 2019b).

The inhibitory activities of *W. indica* essential oil were highly significant. In the 1st to the 3rd hr, edema was significantly reduced ($p < 0.001$) by all the analyzed doses (100, 200 and 400 mg/kg). The *W. indica* essential oil activities at these doses are also equivalent to that of the standard drug used (Ibuprofen). However, at the 4th hr, there were significant reduction in the inhibitory activities of all doses; the 100, 200 and 400 mg/kg exhibited $p < 0.5$, non-significant (ns) and $p < 0.01$ respectively. This show that prolonged exposure to the essential oils of *W. indica* beyond the 3rd hr has no significant activity on the inflammation. The results confirmed earlier findings that *W. indica* extract exhibits a noticeable anti-inflammatory effect in accordance with the traditional uses of the plant. In the present study, oral treatment with *W. indica* essential oil markedly inhibited carrageenan-induced paw oedema in rats in a dose and time dependent property. This treatment steadily attenuated the paw oedema induced by carrageenan, as well as by numerous inflammatory mediators participating in the carrageenan-induced inflammation such as bradykinin, histamine, substance P and platelet-activating factor (Gilligan et al. 1994). This evidence suggests that the anti-inflammatory actions of the essential oil of *W. indica* are related to the inhibition of one or more inflammation mediator pathways involved in the effects of these mediators (Avoseh et al. 2019b).

It should be noted that the two major components of the oil namely limonene and sabinene which are reported to have anti-inflammatory properties under cell culture systems. Limonene has been identified as an anti-inflammatory and anti-tumor candidate. In an earlier study, at low concentration of limonene (7.34 mM), it decreased the production of ROS in eotaxin-stimulated HL-60 clone 15 cells. In the same study, NF-kappa B formation, chemotaxis and eosinophil migration were significantly reduced by limonene at 14.68 mmol/L (Hirota et al. 2010). Limonene has also been reported to be an effective inhibitor of lipopolysaccharide (LPS)-induced NO and prostaglandin E2 production in RAW 264.7 cells. In addition, other cytokines (TNF-α, IL-1β, and IL-6) were also suppressed (Yoon et al. 2010). Sabinene showed an anti-inflammatory activity by inhibiting nitric oxide production in lipopolysaccharide (LPS) triggered (Valente et al. 2013, Kumar et al. 2014). Therefore, the mechanism of pain and anti-inflammatory of *W. indica* essential oil could be attributed to the ability of the major constituents to inhibit several pain mediators synthesized during the study. However, it should be considered that minor and major components, as well as synergistic interactions amongst the substances, could contribute to the studied pharmacological properties.

Moringa oleifera Lam

Essential oils from leaves, flowers, seeds, barks, and roots of *Moringa oleifera* were hydrodistilled and obtained in yields of 0.3% in the seed to 3.6% in the leaves. Eicosane was the major component (20.93, 17.12 and 21.59%) of the oils from the flowers, leaves and seeds respectively, while the oils from the barks and roots had naphthalene (18.40%) and benzene isothiocyanato methyl (35.83%) as the main components. In the study, inhibitory actions of the essential oils on denaturation of BSA and rat erythrocyte membrane hemolysis were used to evaluate the anti-inflammatory activity. Agents that could prevent protein condensation, precipitation of denatured protein aggregates and inhibit the COX enzyme or protect the lysosomal membrane from breakdown are useful in attenuating inflammation and its attendant diseases. Essential oils from the different parts of *M. oleifera* ($p < 0.05$) protected the erythrocyte membrane against hemolysis induced by the hypotonic solution in the order root oil>leaf oil>flower oil>seed oil>bark oil>Diclofenac. The ability of the essential oils to maintain erythrocyte membrane integrity was comparable and even higher than that of Diclofenac the standard drug (Otunola and Afolayan 2018). All the oils, exhibited inhibition against protein denaturation at $IC_{50} < 0.1$ mg/mL, except the leaf oil which showed a higher IC_{50} at 0.2 mg/mL. The membrane-stabilizing action of the essential oils followed the same trend as observed for protein denaturation. Erythrocyte membrane stability was enhanced in a concentration-dependent manner, with the oil from the roots exhibiting the highest inhibition (77.50%) of membrane hemolysis at IC_{50}: 0.1 mg/mL (Otunola and Afolayan 2018).

Dennettia tripetala G. Baker

The analgesic and anti-inflammatory activity of the essential oils of the fruits of *Dennettia tripetala* in rodents was evaluated. The analgesic effects of the oil were assessed in mice using the hot plate, acetic acid-induced writhings and formalin test, while carrageenan-induced paw oedema was used to study the anti-inflammatory effects in rats. The essential oil at 25–50 mg/kg exhibited significant ($p < 0.05$) antinociceptive effects comparable to a potent opioid agonist, morphine (10 mg/kg) and non-steroidal anti-inflammatory drugs such as, aspirin (100 mg/kg) and indomethacin (80 mg/kg). The anti-nociceptive effect of the essential oil was also blocked by naloxone (2 mg/kg) in all the models used. The essential oil demonstrated significant ($p < 0.05$) anti-inflammatory effect in the carrageenan-induced paw oedema model of inflammation that is also comparable to dexamethasone (1 mg/kg). The results showed that the essential oil of *D. tripetala* possesses significant anti-nociceptive and anti-inflammatory effects in the animal models used. The results also suggest that the analgesic effects may be mediated both centrally as well as peripherally, while the anti-inflammatory activity may be effective in both early and late phases of inflammation. The results obtained may therefore be used to rationalize the use of the plant in the treatment of pain and fever in traditional medicine (Oyemitan et al. 2008).

Conclusion

The present study indicated that the essential oils from some Nigerian plants exhibited dose and time dependent activity on the anti-nociceptive and the anti-inflammatory activities. Anti-inflammatory activity was shown to mostly result in the suppression of both the prostaglandins and the histamines at certain doses. In addition, the essential oils displayed an opioid-antagonist property typical of NSAIDs drugs. The presence of several terpene constituents of the essential oils acting in synergy were suggested as agents causing the effect. The study therefore confirmed the ethno pharmacological basis of the use of these plants in traditional medicine for treating pain, swellings, rheumatoid arthritis, etc. Further study on the mechanism of pro-inflammatory inhibition will provide a further basis for this application.

References

Aboluwodi, A.S., N.O. Avoseh, A.O. Lawal, I.A. Ogunwande and A.A. Giwa. 2017. Chemical constituents and anti-inflammatory activity of essential oils of *Datura stramonium* L. J. Med. Plant Stud. 5(1): 21–25.

Adeosun, T.E., N.O. Avoseh, I.A. Ogunwande, I.P. Raji and A.O. Lawal. 2017. Composition and anti-inflammatory activity of essential oil of *Jatropha curcas* L. Nat Prod Commun. 12(3): 439–440.

Amir, F., C.K. Wong, I.M. Eldeen, Z.M. Asmawi and H. Osman. 2012. Volatile constituents and bioactivities of *Mimusops elengi* Flowers. Lat. Am. J. Pharm. 31(2): 331–335.

Avoseh, N.O., O.O. Oyedeji, K. Aremu, B.N. Nkeh-Chungag, S.P. Songca, S.O. Oluwafemi, et al. 2015. Chemical composition and anti-inflammatory activities of the essential oils from *Acacia mearnsii* de wild. Nat. Prod. Res. 29(12): 1184–1188.

Avoseh, N.O., I.A. Ogunwande, Y. Yahaya and P.O. Afolabi. 2019a. Root bark of *Cordia millenii* essential oil: anti-inflammatory and anti-nociceptive activities. Eur. J. Med. Plants. 30(3): 1–7.

Avoseh, N.O., I.A. Ogunwande, O.A. Lawal, J. Atabo, A. Roberta and F. Guido. 2019b. Anti-inflammatory and anti-nociceptive activities of essential oil of *Waltherica indica*. Bol. Latinoam. Caribe Plant Med. Aromá. 18(6): 566–576.

Avoseh, N.O., A.O. Lawal, I.A. Ogunwande, R. Ascrizzi, F. Guido and E. Amoo. 2020a. *In vivo* anti-inflammatory and anti-nociceptive activities, and chemical constituents of essential oil from the leaf of *Gardenia jasminoides* J. Ellis (Rubiaceae). Trends. Phytochem. Res. 4(4): 203–212.

Avoseh, N.O., I.A. Ogunwande, G.O. Ojenike and F.M. Mtunzi. 2020b Volatile composition, toxicity, analgesic and anti-inflammatory activities of *Mucuna pruriens*. Nat. Prod. Commun. 15(7): 1–9.

Avoseh, N.O., F.M. Mtunzi, I.A. Ogunwande, A.D. Shobande, J.E. Nduka, R. Ascrizzi, et al. 2021. *Albizia lebbeck* and *Albizia zygia* volatile oils exhibit antinocicpetive and anti-inflammatory properties in pain models. J. Ethnopharmacol. 268(2021): 113676. https://doi.org/10.1016/j.jep.2020.113676.

Avoseh, N.O., I.A. Ogunwande and H.O. Oshikoya. 2022. Essential oil from the stem bark of *Casuarina equisetifolia* exerts anti-inflammatory and anti-nociceptive activities in rats. Braz J Pharm Sci. 58: 1-11.

Belsito, D., D. Bickers, M. Bruze, P. Calow, H. Greim, J.M. Hanifin, et al. 2007. A toxicologic and dermatologic assessment of ionones when used as fragrance ingredients. Food Chem. Toxicol. 45: S130–S167. https://doi.org/10.1016/j.fct.2007.09.067.

Brito, R.G., A.G. Guimarães, J.S. Quintans, M.R. Santos, D.P. de Sousa, D. Jr. Badaue-Passos, et al. 2012. Citronellol, a monoterpene alcohol, reduces nociceptive and inflammatory activities in rodents. J. Nat. Med. 66(4): 637–44.

Carretero, M.E., J.L. López-Pérez, M.J. Abad, P. Bermejo, S. Tillet, A. Israel, et al. 2008. Preliminary study of the anti-inflammatory activity of hexane extract and fractions from *Bursera simaruba* (Linn) Sarg. (Burseraceae) leaves. J. Ethnopharmacol. 116: 11–15.

Cho, J.Y., H.J. Chang, S.K. Lee, H.J. Kim, J.K. Hwang and Chun HS. 2007. Amelioration of dextran sulfate socium-induced colitis in mice by oral administration of β-caryophyllene, a sesquiterpene. Life Sci. 80: 932–939.

de las Heras, B. and S. Hortelano. 2009. Molecular basis of the anti-inflammatory effects of terpenoids. Inflamm. Allergy Drug Targets. 8(1): 28–39.

Ferreira, A.A., F.A. Amaral, D.G. Duarte, P.M. Oliveira, R.B. Alves, D. Silveira, et al. 2006. Antinociceptive effect from *Ipomoea cairica* extract. J. Ethnopharmacol. 105(1–2): 148–153.

Gilligan, J.P., J.S. Lovato, M.D. Erion and A.Y. Jeng. 1994. Modulation of carrageenan-induced hind paw edema by substance P. Inflammation. 18: 285–292.

Harrison, E.H. and L. Quadro. 2018. Apocarotenoids: emerging roles in mammals. Annu. Rev. Nutr. 38: 153–172. https://doi.org/10.1146/annurev-nutr-082117–051841.

Hart, P.H., C. Brand, C.F. Carson, T.V. Riley, R.H. Prager and JJ. Finlay-Jones. 2000. Terpinen-4-ol, the main component of the essential oil of *Melaleuca alternifolia* (tea tree oil), suppresses inflammatory mediator production by activated human monocytes. Inflamm. Res. 49(11): 619–626.

Hirota, R., R.N. Ngatu, H. Nakamura and M. Sawamura 2010. Anti-inflammatory effects of limonene from Yuzu (*Citrus junos* Tanaka) essential oil on eosinophils. J Food Sci. 75: 87–91.

Jeena, K., V.B. Liju and R. Kuttanract. 2013. Antioxidant, anti-inflammatory and antinociceptive activities of essential oil from ginger. Indian J. Physiol. Pharmacol. 57(1): 51–62.

Juergens, U.R., T. Engelen, K. Racké, M. Stöber, A. Gillissen and H. Vetter 2004. Inhibitory activity of 1,8-cineole (eucalyptol) on cytokine production in cultured human lymphocytes and monocytes. Pulm. Pharmacol. Ther. 2004. 17(5): 281–287.

Kang, C.H., R.G.P.T. Jayasooriya, Y.H. Choi, S.K. Moon, W.J. Kim and G.Y. Kim. 2013. β-Ionone attenuates LPS-induced pro-inflammatory mediators such as NO, PGE2 and TNF-α in BV2 microglial cells via suppression of the NF-κB and MAPK pathway. Toxicol. Vitro. 27: 782–787. https://doi.org/10.1016/j.tiv.2012.12.012.

Ku, C.M. and J.Y. Lin. 2015. Farnesol, a sesquiterpene alcohol in herbal plants, exerts anti-Inflammatory and antiallergic effects on ovalbumin-sensitized and–challenged asthmatic mice. Evid Based Compleme Alter. Med. 2015: 1–12. doi: org/10.1155/2015/387357.

Kumar, R.A., N. Ajikumar, R.B. Rameshkumar and S. Appian. 2014. The essential oil constituents of *Zornia diphylla* (L.) Pers, and anti-inflammatory and antimicrobial activities of the oil. Rec. Nat. Prod. 8: 385–393.

Lawal, O.A., N.O. Avoseh, J.O. Tinubu, I.A. Ogunwande, R. Ascrizzi and G. Flamini. 2020. Anti-nociceptive property, anti-inflammatory activity and constituents of essential oils from the leaves and stem bark of *Turnera diffusa* Wild (Turneraceae) grown in Nigeria. J. Biol. Active Prod. Nature. 10(6): 473–483.

Li, J., Y. Yin, L. Wang, P. Liang, M.H. Li, X. Liu, et al. 2016. Synthesis, characterization, and anti-inflammatory activities of methyl salicylate derivatives bearing piperazine moiety. Molecules. 21(11): 1544–1548.

Li, J., R. Thangaiyan, K. Govindasamy and J.X. Wei. 2020. Anti-inflammatory and anti-apoptotic effect of zingiberene on isoproterenol-induced myocardial infarction in experimental animals. Hum. Exp. Toxicol. 39(11): 1–13.

Liao, P.C., T.S. Yang, J.C. Chou, J. Chen, S.C. Lee, Y.S. Kuo, et al. 2015. Anti-inflammatory activity of neral and geranial isolated from fruits of *Litsea cubeba* Lour. J. Funct. Foods. 19(Part A): 248–258.

Mai, M., K. Isoko, N. Hiroto, M. Hironari, S. Keiichiro, K. Yuko, et al. 2014. Suppression of allergic and inflammatory responses by essential oils derived from herbal plants and citrus fruits. Int. J. Mol. Med. 33(6): 1643–1651.

Nishijima, C.M., E.G. Ganev, L. Mazzardo-Martins, D.F. Martins, L.R.M. Rocha, A.R.S. Santos, et al. 2014. Citral: A monoterpene with prophylactic and therapeutic anti-nociceptive effects in experimental models of acute and chronic pain. Eur. J. Pharmacol. 736: 16–25.

Ogunwande, I.A., O.N. Avoseh, K.N. Olasunkanmi, O.A. Lawal, R. Ascrizzi and G. Flamini. 2019a. Chemical composition, antinociceptive and anti-inflammatory activities of essential oil of *Bougainvillea glabra*. J. Ethnopharmacol. 232: 188–192.

Ogunwande, I.A., N.O. Avoseh, D.A. Igile, O.A. Lawal, R. Ascrizzi and F. Guido. 2019b. Chemical constituents, anti- nociceptive and anti-inflammatory and activities of essential oil of *Phyllathus muellerianus* (O. Kuntze) Excell. Nat. Prod. Commun. 14(5): 1–7.

Oluwa, A.A., N.O. Avoseh, O. Omikorede, I.A. Ogunwande and A.O. Lawal. 2017. Study on the chemical constituents and anti-inflammatory activity of essential oil of *Petiveria alliacea* L. J. Pharm. Res. Int. 15(1): 1–8. doi: 10.9734/BJPR/2017/31331.

Otunola, G.A. and A.J. Afolayan. 2018. Chemical composition, antibacterial and *in vitro* anti-inflammatory potentials of essential oils from different plant parts of *Moringa oleifera* Lam. Ame. J. Biochem. Biotechnol. 14(3): 210.220. doi: 10.3844/ajbbsp.2018.210.220.

Oyemitan, I.A., E.O. Iwalewa, M.A. Akanmu and T.A. Olugbade. 2008. Antinociceptive and anti-inflammatory effects of essential oil of *Dennettia tripetela* G. Baker (Annonaceae) in rodents. Afr. J. Trad. Complement Altern. Med. 5(4): 355-362.

Park, S.Y., M.L. Jin, Y.H. Kim, Y. Kim and S.J. Lee. 2012a. Anti-inflammatory effects of aromatic-turmerone through blocking of NF-kappaB, JNK, and p38 MAPK signaling pathways in amyloid beta-stimulated microglia. Int. Immunopharmacol. 14(1): 13–20.

Park, S.Y., Y.H. Kim, Y. Kim and S.J. Lee. 2012b. Aromatic-turmerone's anti-inflammatory effects in microglial cells are mediated by protein kinase A and heme oxygenase-1 signaling. Neurochem. Int. 61: 767–777.

Passos G.F., E.S. Fernandes, F.M. da Cunha, J. Ferreira, L.F. Pianowski, M.M. Campos, et al. 2007. Anti-inflammatory and anti-allergic properties of the essential oil and active compounds from *Cordia verbenacea*. J. Ethnopharmacol. 110(2): 323–333.

Piccinelli, A.C., J.A. Santos, E.C. Konkiewitz, S.A. Oesterreich, A.S.N. Formagio, J. Croda, et al. 2015. Antihyperalgesic and antidepressive actions of (*R*)-(+)-limonene, α-phellandrene, and essential oil from *Schinus terebinthifolius* fruits in a neuropathic pain model. Nutr. Neurosci. 18(5): 217–224.

Ritter, A.M., F.Q. Ames, F. Otani, R.M.W. de Oliveira, R.K.N. Cuman and C.A. Bersani-Amado. 2014. Effects of anethole in nociception experimental models. Evid. Based Complem. Altern. Med. 2014: 345829. https://doi.org/10.1155/2014/345829.

Saad, N.Y., C.D. Muller and A. Lobstein. 2013. Major bioactivities and mechanism of action of essential oils and their components. Flav. Fragr. J. 28: 269–279. https:// doi.org/10.1002/ffj.3165.

Salud, P.G., Z.S. Miguel, A.G. Lucina and R.L. Miguel. 2011. Anti-inflammatory activity of some essential oils. J. Essent. Oil Res. 23(5): 38–44.

Shah, A.J., A.H. Gilani, K. Abbas, M. Rasheed, A. Ahmed and V.U. Ahmad. 2011. Studies on the chemical composition and possible mechanisms underlying the antispasmodic and bronchodilatory activities of the essential oil of *Artemisia maritima* L. Arch. Pharm. Res. (Seoul). 34: 1227–1238. https://doi.org/10.1007/s12272- 11-0801-0.

Silva, R.O., F.B. Sousa, S.R. Damasceno, N.S. Carvalho, V.G. Silva, F.R. Oliveira, et al. 2014. Phytol, a diterpene alcohol, inhibits the inflammatory response by reducing cytokine production and oxidative stress. Fundam. Clin. Pharmacol. 28(4): 455–64.

Tang, F., F. Chen, X. Ling, Y. Huang, X. Zheng, Q. Tang, et al. 2015. Inhibitory effect of methyleugenol on IgE-diated allergic inflammation in RBL-2H3 cells. Mediat. Inflamm. 2015: 1–10. https://doi.org/10.1155/2015/463530.

Valente, J., M. Zuzarte, M.J. Goncalves, M.C. Lopes, C.S.L. Cavaleiro, L. Salqueiro, et al. 2013. Antifungal, anti- oxidant and anti-inflammatory activities of *Oenanthe crocata* L. essential oil. Food Chem. Toxicol. 62: 349–354.

Vijayalaxmi, A., B. Vasudha and B. Nazia. 2015. Anti-arthritic and anti-inflammatory activity of beta-caryophyllene against Freund's complete adjuvant induced arthritis in Wistar rats. J. Bone. Res. Rep. 1: 1–10.

Wang, Z.J., B. Tabakoff, S.R. Levinson and T. Heinbockel. 2015. Inhibition of Nav 1.7 channels by methyl eugenol as a mechanism underlying its antinociceptive and anesthetic actions. Acta Pharmacol. Sin. 36: 791–799. https://doi.org/10.1038/ aps.2015.26.

Xin, W., C. Huang, X. Zhang, S. Xin, Y. Zhou, X.W. Ma, et al. 2014. Methyl salicylate lactoside inhibits inflammatory response of fibroblast-like synoviocytes and joint destruction in collagen-induced arthritis in mice. Brit. J. Pharmacol. 171(14): 3526–3538.

Yang, S., J. Liu, L. Jiao and L. Jiao. 2020. Ar-turmerone exerts anti-proliferative and anti-inflammatory activities in HaCaT keratinocytes by inactivating hedgehog pathway. Inflammation. 43: 478–486. https://doi.org/10.1007/s10753-019-01131-w.

Yoon, W., N.H. Lee. and C. Hyun. 2010. Limonene suppresses lipopolysaccharide-induced production of nitric oxide, prostaglandin E2, and pro-inflammatory cytokines in RAW 264, Macrophages. J. Oleo Sci. 421: 415–421. https://doi.org/10.5650/jos.59.415.

10

Antiparasitic, Antibacterial and Antifungal Properties of Essential Oils Obtained from Medicinal Plants Available in the Southeast Asian Region

Zainul Amiruddin Zakaria[1*], *Fezah Othman*[2], *Mohd. Hijaz Mohd. Sani*[1], *Noraina Muhamad Zakuan*[2], *Wan Salman Wan Saudi*[1] *and Tavamani Balan*[3]

Introduction

Infectious Diseases

Infectious diseases represent one-fourth of the global concern of disease and remain a major cause of morbidity and mortality in many countries, particularly developing countries, with people being affected and suffering heavily from the burden of ill health and death, particularly in infants and children (Gulati et al. 2021). These infectious diseases are mainly treated with anti-infective drugs (i.e., antibiotics or antihelmintic) or antiparasitic drugs. Despite the considerable advances in the anti-infective management of the immunosuppressed host that was achieved in the last 20 yr, which include the development of antibiotics with a broad-spectrum and high bactericidal activity, infection still remains a major cause of death in compromised hosts and a number of limitations of therapy persist.

Hence, there are urgent and continuous needs to discover new anti-infective compounds with novel mechanisms of action and diverse chemical structures

[1] Borneo Research on Algesia, Inflammation and Neurodegeneration (BRAIN) Group, Department of Biomedical Sciences, Faculty of Medicine and Health Sciences, Universiti Malaysia Sabah, Jalan UMS, 88400, Kota Kinabalu, Sabah, Malaysia.
[2] Department of Biomedical Sciences, Faculty of Medicine and Health Sciences, Universiti Putra Malaysia, 43400, UPM Serdang, Selangor, Malaysia.
[3] Faculty of Pharmacy and Health Sciences, Royal College of Medicine Perak, Universiti Kuala Lumpur, 30450 Ipoh, Perak, Malaysia.
* Corresponding author: zaz@ums.edu.my

to cater for the re-emerging and new infectious diseases (Zinner 2005). The unlimited prospects of new drugs particularly those with anti-infective properties, from a wide number of plant-derived secondary metabolites make them a potential reservoir worth exploring (Mustafa et al. 2017).

Medicinal Plants as a Source of Medicinal Agents

Plants are an essential, but often underestimated health resource with many applications to humans, especially in the prevention and management of lifestyle-related chronic diseases, and in meeting the health needs of the ageing populations. Indigenous people in various parts of the world have established a unique approach towards their traditional system of medicine of which medicinal plants are an important element and among the most renowned are the Chinese, Indian and Arabs systems of medicine (Petrovska 2012, Karunamoorthi et al. 2013). Currently, there is a rampant increase in the use of medicinal plants throughout the world as an alternative treatment to various diseases (Astutik et al. 2019).

The Southeast Asia Region (SeAR), especially, is a region of high plant diversity with an estimated 50,000 flowering plant species (Middleton et al. 2019). People in SeAR, which consists of 12 member states, including Malaysia, have a long history of using traditional medicine, particularly, for a wide range of health conditions, including as anti-infective agents against the infection of bacteria, fungal and parasites. Aromatic plants, in particular have a very special place among the different ethnic groups living in SeAR. They are a group of medicinal plants that are found in abundance in tropical and sub-tropical Asia due to their varied and suitable ecological conditions (Chomchalow 2002). This could help explain why the application of aromatic plants by Asian people in many traditional ways have been well-documented since the prehistoric era. Even in the past 20 yr, as reported by Oyen and Nguyen (1999) have highlighted SeAR as a major player in the production of some essential oils, including cajeput oil, citronella oil, cananga oil, lemongrass oil, to name a few.

Essentials Oils as Potential Sources of Anti-infective Drugs

Aromatic plants have been traditionally used not only as food ingredients but also to treat a plethora of ailments. Various parts of aromatic plants can be used to treat, control and manage a variety of microbial (Kebede et al. 2021) and parasitic (Wink 2012) infections. One of the disreputable ways of using aromatic plants therapeutically is by exposing the plant parts involved, which can either be the seeds, roots, berries, leaves, flowers or barks, to steam distillation, a method widely used to obtain the Essential Oils (EOs) (Wińska et al. 2019, Dawood et al. 2021). Essential oils are mixtures of natural volatile, but complex compounds characterized by a strong odour and are formed by aromatic plants as secondary metabolites (Alviano and Alviano 2009). The oils consist mainly of a balance mixture of monoterpenes, sesquiterpenes, and their oxygenated derivatives (alcohols, aldehydes, esters, ethers, ketones, phenols, and oxides) and have been known since antiquity to possess therapeutic effects, including for the treatment of microbial and parasitic infections. Moreover, the curiosity about the antimicrobial potentials of essential oil (EO) is growing because of the broad range of activities, natural origins, and Generally Recognized As Safe (GRAS) status of EOs.

In the next part of this chapter, all the reports related to the ability of EOs to exert antibacterial and antifungal activities focusing only those obtained from plants that grow in the SeAR in the past 3 yr (2019–2021) are highlighted. Additional reviews on the efficacy of EOs against parasitic infection at various stages of infection are also made.

Antiparasitic Activity of Essential Oils

In this antiparasitic section, a description was presented based on the type of parasite under investigation to give clear information to the readers and a summary on the findings related to antiparasitic studies of medicinal plants obtained from SeAR is given in Table 10.1.

Anti-Giardia Duodenalis Activity

Pintong et al. (2020) studied the ability of EO of *Ageratum conyzoides* against *Giardia duodenalis* trophozoites. A. conyzoides was collected from the district of Chiang Khong in Chiang Rai, Thailand during March 2016–2017 and only the fresh Leaves of White-Purple Flowers (LWPF) and fresh Flowers of Purple-Flowers (FPF) were subjected to hydrodistillation to obtain the respective EO. The EOs of LWPF and FPF were subjected to the *in vitro G. duodenalis* trophozoites and also analyzed to determine their phytoconstituents using the GC-MS. The results show that both EOs demonstrated highly active anthelmintic activity against *G. duodenalis* indicated by the IC_{50} value of approximately 35.0 and 89.3 μg/mL, respectively. However, the IC_{50} value of the reference drug, metronidazole, was not provided in the report. With regards to the phytoconstituents of both EOs, LWPF contained 32 while FPF contained 35 phytochemicals content. Based on the analyses, several classes of compounds, namely, chromene group (61.0%), sesquiterpenes (26.9%) and monoterpenes (0.2%) were detected in the EO of LWPF, whereas the EO of FPF contained the chromene group (57.5%), sesquiterpenes (29.6%) and monoterpenes (1.2%). Further analyses revealed that the primary phytoconstituents contain the chromene group (precocene I, precocene II, and 6-vinyl-7-methoxy-2,2-dimethylchromene); sesquiterpenes (i.e., β-caryophyllene, α-caryophyllene, germacrene D, copaene, caryophyllene oxide, and β-bourbonene), and; monotepenes (α-pinene, camphene, β-pinene, limonene, and endo-bornyl acetate).

Anti-Trypanosoma Brucei Brucei Activity

Le et al. (2019) studied the anti-trypanosomal potential against *Tbb* of 37 EOs that were extracted from various parts of Vietnamese plants. The plants' part(s) were collected between November 2014 and August 2015 followed by hydrodistillation to yield the respective EO. In the primary anti-trypanosomal screening, 33 EOs were tested at concentrations of 50 nL/mL and 25 nL/mL, while four EOs were

Table 10.1 Summary of phytoconstituents of EOs from various plants obtained from SeAR that exerted antiparasitic activity

Parasite	Plant	Family	Parts	Location	Phytoconstituents	References
Giardia duodenalis	*Ageratum conyzoides* L.	Asteraceae	Leaves of white-purple flowers (LWPF) Fresh flowers of purple-flowers (FPF)	Chiang Khong district, Chiang Rai, Thailand	• EO of LWPF contained chromene group (61.0%), sesquiterpenes (26.9%) and monoterpenes (0.2%) • EO of FPF chromene group (57.5%), sesquiterpenes (29.6%) and monoterpenes (1.2%). • Primary phytoconstituents detected in both EO included chromene group (precocene I, precocene II, and 6-vinyl-7-methoxy-2,2-dimethylchromene); sesquiterpenes (i.e. β-caryophyllene, α-caryophyllene, germacrene D, copaene, caryophyllene oxide, and β-bourbonene), and; monotepenes (α-pinene, camphene, β-pinene, limonene, and endo-bornyl acetate).	Pintong et al. 2020
Trypanosoma brucei brucei (Tbb)	*Alpinia galanga* (L.) Willd.	Zingiberaceae	Rhizomes	Vietnam (various area; specific location not provided)	• GC-MS and GC-FID analyzes of the effective EOs revealed: • Sesquiterpenes were the main active component in the three EOs belonging to C. longa, C. zedoaria, and Z. officinale • Monoterpenes were the main active component of L. cuceba EO. • Major monoterpenes identified were citronellal (43.1%), isopulegol (11.1%), limonene (8.7%), pulegol (6.5%), linalool (5.6%), and citronellol (5.2%)	Le et al. 2019

Table 10.1 contd....

...Table 10.1 contd.

Parasite	Plant	Family	Parts	Location	Phytoconstituents	References
					• Major sesquiterpenes in both EOs belonging to *C. longa* and *Z. officinale* were α-zingiberene, β-bisabolene, β-sesquiphellandrene, and *ar*-curcumene with the percentages of 25.4, 3.4, 18.3, and 5.2% and, 27.7, 7.3, 8.1, and 2.7%, respectively. • Other major components of EO of *C. longa* includes oxygenated sesquiterpenes (i.e., α-turmerone (10.3%), germacrone (3.3%), curlone (5.2%), and *ar*-turmerone (9.9%). • Other major components of EO of *Z. officinale* includes monoterpenes (i.e. β-phellandrene (14.8%) and camphene (6.9%)). • In addition, major components of *C. zedoaria* EO include 8,9-dehydro-9-formyl cycloisolongifolene (29.3%), curdione (13.5%), and germacrone (9.0%). • In the first fractionation process yielded CF1 with sesquiterpenes and CF5 with oxygenated sesquiterpenes. • In the second fractionation process both fractions yielded three pure compounds, β-sesquiphellandrene, *ar*-curcumene, and curlone.	

Parasite	Plant	Family	Parts	Location	Phytoconstituents	References
	Amomum aromaticum Roxb.	Zingiberaceae	Fruits			
	Amomum schmidtii (K. Schum.) Gagnep.	Zingiberaceae	Rhizomes			
	Anethum graveolens L.	Apiaceae	Fruits			
	Artemisia annua L.	Asteraceae	Leaves			
	Blumea lanceolaria (Roxb.) Druce	Asteraceae	Leaves			
	Cinnamomum cassia (L.) J. Presl	Lauraceae	Stem barks			
	Clausena indica (Dalzell) Oliv.	Rutaceae	Leaves			
	Coriandrum sativum L.	Apiaceae	Fruits			
	Curcuma longa L.	Zingiberaceae	Rhizomes			
	Curcuma zedoaria (Christm.) Roscoe	Zingiberaceae	Rhizomes			
	Dysphania ambrosioides (L.) Mosyakin & Clemant	Amaranthaceae	Leaves and fruits			
	Elsholtzia blanda (Benth.) Benth.	Lamiaceae	Leaves			
	Elsholtzia ciliata (Thunb.) Hyl.	Lamiaceae	Leaves			
	Elsholtzia communis (Collett & Hemsl.) Diels	Lamiaceae	Leaves			
	Elsholtzia penduliflora W. Sm.	Lamiaceae	Leaves			
	Eucalyptus camaldulensis Dehnh.	Myrtaceae	Leaves			

Table 10.1 contd....

...Table 10.1 contd.

Parasite	Plant	Family	Parts	Location	Phytoconstituents	References
	Hedychium coronarium J. Koenig	Zingiberaceae	Rhizomes			
	Hyptis suaveolens (L.) Poit.	Lamiaceae	Leaves			
	Illicium verum Hook. f.	Schisandraceae	Fruits			
	Kaempferia galanga L.	Zingiberaceae	Rhizomes			
	Litsea cubeba (Lour.) Pers.	Lauraceae	Fruits			
	Litsea cubeba (Lour.) Pers.	Lauraceae	Leaves			
	Melaleuca alternifolia (Maiden & Betch) Cheel	Myrtaceae	Leaves			
	Melaleuca cajuputi Powell	Myrtaceae	Leaves			
	Ocimum gratissimum L.	Lamiaceae	Leaves			
	Ocimum tenuiflorum L.	Lamiaceae	Leaves			
	Piper sarmentosum Roxb.	Piperaceae	Leaves			
	Platycladus orientalis (L.) Franco	Cupressaceae	Leaves			
	Plectranthus amboinicus (Lour.) Spreng.	Lamiaceae	Leaves			
	Pluchea indica (L.) Less.	Asteraceae	Leaves			
	Pogostemon cablin (Blanco) Benth.	Lamiaceae	Leaves			
	Vitex trifolia L.	Lamiaceae	Leaves			
	Zingiber montanum (J.Koenig) Link ex A.Dietr.	Zingiberaceae	Rhizomes			
	Zingiber officinale Roscoe	Zingiberaceae	Rhizomes			
	Zingiber zerumbet (L.) Roscoe ex Sm.	Zingiberaceae	Rhizomes			

Parasite	Plant	Family	Parts	Location	Phytoconstituents	References
Rhipicephalus (Boophilus) microplus (Acari: *Ixodidae*)	*Citrus hystrix* DC.	Rutaceae	Part of each plant used was not given	EOs of both plants were purchased from a local company (Tropical Bioessence Sdn. Bhd., Malaysia)	• Several medically importance phyto-constituents such as: • *C. hystrix* EO contained α-myrcene, linalool, α-pinene, citronellal, neral and citronellol • *C. citratus* EO contained linalool, eucalyptol, neral, camphene, α-pinene and citronellal • Combination of both plants EO contained most of the above-mentioned compounds plus two new phytoconstituents, namely 2,4-undecadien1-ol and 1,7-nonadien-4-ol,4,8-dimethyl-.	Shezryna et al. (2021)
	Cymbopogon citratus (DC.) Stapf.	Poaceae				

NA – Not available; ND – Not determined

tested as at 10 nL/mL and 5 nL/mL, with suramin used as the reference drug. Only EOs that showed more than 50% inhibition of the parasite growth at 25 or 10 nL/mL were subjected to the IC_{50} determination. In this case the selected EOs were tested at the concentration range of 50–0.02 nL/mL, except for the EO of *Cinnamomum cassia*, *Curcuma zedoaria*, and *Z. zerumbet*, which were tested at the concentration range of 10–0.005 nL/mL, and EO of *Dysphania ambrosioides* that was tested at the concentration range of 5–0.002 nL/mL. Other than that, the EOs were also subjected to the cytotoxicity study and phytoconstituents analyses using GC-MS and GC-FID as described by Le et al. (2017). However, in this study the concentrations of EOs tested for cytotoxicity was in the range of 50 to 1.40 nL/mL, instead of 50 to 0.02 nL/mL. In addition, the authors also carried out isolation of the bioactive compound from the most effective EO using the column chromatography technique. From the results obtained, 19 out of 37 EOs showed promising anti-trypanosomal activity against *Tbb,* whereby less than 3% of viable parasites were recorded at the lowest tested concentration. Based on the criteria set in earlier research (IC_{50} < 2 µg/mL: strongly effective; IC_{50} between 2 and 20 µg/mL: moderately active) it is suggested that many of these EOs have moderate anti-trypanosomal activity that required further in-depth study. Of the 19 EOs, only three samples belonging to the plants from *Zingiberaceae* family, namely *C. longa*, *C. zedoaria*, and *Z. officinale*, and one sample belonging to the *Lauraceae* family (i.e., *Litsea cubeba*) demonstrated the most effective anti-trypanosomal activity indicated by the lowest IC_{50} value of approximately 3.20, 2.50, 3.10, and 2.70 nL/mL, respectively. Interestingly, these EOs also possessed the SI value of more than 10 (SI > 10) compared to others. On GC-MS and GC-FID analyses of these active EOs, more than 85% of each EO composition were characterized. Sesquiterpenes were found to be the main active component in the three EOs belonging to the *Zingiberaceae* family, while monoterpenes were identified as the main active component of *L. cuceba* EO. The major monoterpenes identified were citronellal (43.1%), isopulegol (11.1%), limonene (8.7%), pulegol (6.5%), linalool (5.6%), and citronellol (5.2%). On the other hand, α-zingiberene, β-bisabolene, β-sesquiphellandrene, and *ar*-curcumene were identified to be the major sesquiterpenes in both EOs belonging to *C. longa* and *Z. officinale* with the recorded percentages of 25.4, 3.4, 18.3, and 5.2% and, 27.7, 7.3, 8.1, and 2.7%, respectively. Besides that, what made these two EOs different in terms of their composition was that the EO of *C. longa* contained oxygenated sesquiterpenes (i.e., α-turmerone (10.3%), germacrone (3.3%), curlone (5.2%), and *ar*-turmerone (9.9%), whereas the EO of *Z. officinale* contained monoterpenes (i.e., β-phellandrene (14.8%) and camphene (6.9%) as their other major components, respectively. With regards to the EO of *C. zedoaria*, the major components detected were 8,9-dehydro-9-formyl cycloisolongifolene (29.3%), curdione (13.5%), and germacrone (9.0%). Further investigations involving fractionation using column chromatography was performed on the EO of *C. longa* since it was the only EO that produced the highest yield. Using silica gel and n-hexane-ethyl acetate system as eluants in the first process two important groups, CF1 with sesquiterpenes and CF5 with oxygenated sesquiterpenes. In the second fractionation process that used silver nitrate impregnated silica gel, both fractions yielded three pure compounds, β-sesquiphellandrene, *ar*-curcumene, and curlone with a purity of 97.0, 97.4, and 91.7%, respectively. These compounds, together with two commercially available ones, α-zingiberene and *ar*-turmerone, were then tested for anti-trypanosomal activity and cytotoxicity. From the results obtained, only curlone gave the lowest IC_{50} value, which was 1.4 μg/mL (6.3 μM) against the *Tbb* bloodstream form and with the recorded SI that was greater than 10 (SI > 10) compared to cytotoxicity on mammalian cells. On the other hand, the rest of the compounds produced an IC_{50} that was greater and SI value that was lower than curlone in the sequence of α-zingiberene (IC_{50} = 6.9 μg/mL), β-sesquiphellandrene (IC_{50} = 9.9 μg/mL), *ar*-curcumene (IC_{50} = 13.4 μg/mL) and *ar*-turmerone (IC_{50} = 28.8 μg/mL), respectively. Taking the value of IC_{50} and SI into account, a compound like curlone could be considered as a promising drug for the development of a new treatment against Human African trypanosomiasis. In comparison, suramin exerted anti-trypanosomal activity at the recoded IC_{50} value of 21.5 μg/mL.

Anti-Rhipicephalus (Boophilus) Microplus Activity

The EO of *Citrus hystrix* and *Cymbopogon citratus* were studied by Shezryna et al. (2020) for its *in vitro* acaricidal activity against the larvae of cattle tick *Rhipicephalus (Boophilus) microplus* (Acari: *Ixodidae*). In this study, the EOs of *C. hystrix* and *C. citratus* were purchased from a local company (Tropical Bioessence Sdn. Bhd., Malaysia). The EOs were subjected separately and in combination to a larval immersion test against 7 to 14 d old *R. (B.) microplus* larvae to determine their acaricidal activity. Counting for the number of dead and live larvae was done manually after 24 and 48 hr. In addition, the EOs were also subjected to phytoconstituents analysis using the GC-MS. The EO of *C. hystrix* was prepared in 10 concentrations, while the EO of *C. citratus* was prepared in seven concentrations and the EO of the combination of both plants was prepared in nine concentrations, all ranging between 0.50–40.00%. After 24 hr, the recorded LC_{50} value of EOs of *C. hystrix*, *C. citratus* or their combination was approximately 12.0, 1.2 and 1.5% (*v/v*) whereas at 48 hr, the recorded LC_{50} value was approximately 11.0, 1.1 and 1.5% (*v/v*), respectively. On the other hand, the LC_{90} value of the three EOs recorded at 24 hr was 24.8, 6.3 and 2.8% (*v/v*), while at 48 hr the LC_{90} value recorded was 21.7, 6.1 and 2.8% (*v/v*), respectively. Further analysis performed to determine the synergistic factor value from the combination of these EOs at a 1:1 ratio revealed that *C. hystrix* possesses synergistic factor of 7.90 and 7.32 post 24 and 48 hr indicating a synergistic effect, respectively. On the other hand, *C. citratus* synergistic factor of 0.80 and 0.70 was calculated post 24 and 48 hr, respectively. According to the authors, the addition of *C. citratus* into *C. hystrix* resulted in a higher acaricidal effect, while the addition of C. *hystrix* lowers the effect of *C. citratus* indicating an antagonistic effect. Although Ivermectin was mentioned as the reference drug in this study, no data was provided to enable the comparison with each of the EO. In addition, GC-MS analysis of the three EOs revealed the presence of several medically important phytoconstituents such as: (1) EO of

C. hystrix contained α-myrcene, linalool, α-pinene, citronellal, neral and citronellol; (2) EO of *C. citratus* contained linalool, eucalyptol, neral, camphene, α-pinene, and citronellal, and; (3) EO of a combination of both plants contained most of these compounds, in addition to two new phytoconstituents, namely 2,4-undecadien1-ol and 1,7-nonadien-4-ol,4,8-dimethyl.

Antibacterial Activity of Essential Oils

Medicinal plants have been known as one of the sources of lead for antibacterial drugs development and the significant contribution of SeAR in this case is proven and supported through various articles published constantly in the past 20 yr as described below.

Ha et al. (2019), in their attempt to determine the antibacterial activity of Vietnamese plants, have screened the leaf and twig EOs of *Magnolia hypolampra* (Dandy) Figlar against one strain of Gram-positive bacteria (*S. aureus* ATCC13709) and one strain of Gram-negative bacteria (*E. coli* ATCC25922) using the agar disc diffusion method. The wildly cultivated leaves and twigs of *M. hypolampra*, collected in Na Hang Nature Reserve in Thanh Tuong commune, Na Hang district, Tuyen Quang province, North of Vietnam, were subjected to the hydrodistillation process to obtain the respective EO. Each EO was then subjected to the GC-MS analysis followed by the antimicrobial study as briefly described above. Following the antimicrobial screening procedure, the MIC and median inhibitory concentration (IC_{50}) values were also determined using the microdilution broth susceptibility assay in which the selected EOs were tested against three Gram-positive bacteria (i.e., *S. aureus* ATCC13709, *B. subtilis* ATCC6633, and *Lactobacillus fermentum* VTCC N4) and three Gram-negative bacteria (i.e., *S. enterica* VTCC, *E. coli* ATCC25922, and *P. aeruginosa* ATCC15442). The VTCC strains were obtained from Vietnam Type Culture Collection, Vietnam National University, Hanoi. The stock solution of each EO in DMSO was serially diluted (two-fold) in sterile distilled water to prepare the test solutions in a concentration ranging between 16384 to 2 μg/mL. The serial dilutions of the sample without microorganisms represents a negative control, whereas the sterile distilled water and medium serves as a positive control. Following GC-MS analysis, a total of 40 and 41 phytoconstituents, representing 98.9 and 98.5% of the total oil compositions, were identified in the leaf and twig of *M. hypolampra* EOs, respectively. Of the total oil compositions recorded, the major compositions of both EOs were α-pinene (23.7 and 24.4%), β-pinene (36.5 and 41.3%), and germacrene D (14.6 and 5.8%), sabinene (2.9 and 2.8%) and limonene (2.6 and 3.2%), respectively. The antibacterial screening showed that both EOs demonstrated strong antibacterial activity against *E. coli* and *S. aureus* (growth inhibition zones $\geq$ 14.0 mm). The twig EO exerted stronger activity (45.5±2.1 and 45.5±0.7 mm) against the two bacteria in comparison to the leaf EO (30.5±0.7 and 18.5±0.7 mm), respectively. Furthermore, the MIC value recorded for the leaf EO was: (i) 4.1 mg/mL against *S. aureus*, *S. enterica* and *E. coli C. albicans*; (ii) 8.2 mg/mL against *L. fermentum* and *P. aeruginosa*, and; 16.4 mg/mL against *B. subtilis,* whereas the twig EO, which possessed a more effective antibacterial activity, produced the MIC value of: (i) 2.0 mg/mL against *S. aureus*; (ii) 4.1 mg/mL against *S. enterica* and *E. coli*; (iii) 8.2 mg/mL against *B. subtilis*, *L. fermentum* and *P. aeruginosa*. The IC_{50} value recorded further confirmed the effectiveness of the twig EO (1.0–3.4 mg/mL over the leaf EO (18–3.7 mg/mL). The reference drugs, namely ampicillin for Gram-positive bacteria and cefotaxime Gram-negative bacteria, produced the MIC value ranging between 0.004 to 1.2 µg/mL and 0.07 to 19.23 µg/mL, respectively.

Huong et al. (2020a) studied the antibacterial activity of *Zingiber zerumbet* rhizomes EO against two Gram-negative strains (*E. coli* ATCC25922 and *P. aeruginosa* ATCC25923) and four Gram-positive strains (*B. subtilis* ATCC11774, *S. aureus* subsp. *aureus* ATCC11632, *Aspergillus niger* ATCC9763 and *Fusarium oxysporum* ATCC48112). The rhizomes, collected from Bến En National Park, Thanh Hóa Province, Vietnam, was hydrodistilated to obtain rhizome EO, which was then subjected to GC-MS and GC-FID analyses. The rhizome EO was also submitted to an antimicrobial study to determine the MIC value of the said EO against the above microbes using the microdilution broth susceptibility assay. Sterile distilled water and DMSO was used as the positive control, while the reference drug uses was not highlighted. GC-MS and GC-FID analyses of the rhizome EO demonstrated the presence of 27 volatile compounds, which account for 97.1% of the total EO content. Specifically, zerumbone (51.3%), humulene epoxide I (6.4%), humulene epoxide II (5.5%), α-humulene (5.4%) and camphene (4.1%) were identified as the major constituents. The MIC value determination revealed that the rhizome EO exerted inhibitory activity only against *A. niger* (MIC = 50.0±0.5 µg/mL).

Dai et al. (2020) studied the antibacterial activity of EO from different species of *Cinnamomum* (i.e., *Cinnamomum ovatum*, *C. tonkinense*, *C. damhaenensis*, *C. longipetiolatum* and *C. polyadelphum*) obtained from different locations in North Central Vietnam. The leaves and stems of *C. ovatum* and the leaves of *C. tonkinense*, *C. damhaenensis*, *C. longipetiolatum* and *C. polyadelphum* were hydrodistillated to obtain the respective EO, which was then subjected to GC-MS and GC-FID analyses followed by an antibacterial study against three strains of Gram-positive bacteria (i.e., *Enterococcus faecalis* ATCC299212, *S. aureus* ATCC25923 and *B. cereus* ATCC14579) and three strains of Gram-negative bacteria (i.e., *E. coli* ATCC25922, *P. aeruginosa* ATCC2785 and *S. enterica* ATCC13076) using the microdilution broth susceptibility assay to determine the MIC and IC_{50} values of each EO. Each EO underwent a two-fold serial dilution to achieve a concentration range of 16384 to 2 µg/mL before being subjected to the antibacterial assay. A serially-diluted sample without microorganisms serves as a positive control, whereas sterile distilled water and medium served as a negative control. Streptomycin was the reference drug. GC-MS and GC-FID analyses detected 22 and 27 phytoconstituents in the leaf and stem EOs of *C. ovatum*, respectively, whereas 35, 33, 24 and 38 phytoconstitutents were detected in the leaf EO of *C. tonkinense*, *C. damhaenensis*, *C. longipetiolatum,* and *C. polyadelphum*, respectively. Specifically: (i) the leaf EO of *C. ovatum* contained eugenol (71%), eugenyl acetate (10%) and linalool (6%); (ii) the stem EO of *C. ovatum* contained eugenol (71%), linalool (8%), eugenyl acetate (9%);

(iii) the leaf EO of *C. tonkinense* contained linalool (32%), β-phelandrene (23%) and 1,8-cineol (10%) (iv) the leaf EO of *C. damhaenensis* contained linalool (45%), β-selinene (19%) and selin-11-en-4α-ol (7%); (v) the leaf EO of *C. longipetiolatum* contained linalool (76%), hotrienol (3%) and cis-linalool oxide (3%), and (vi) the leaf EO of *C. polyadelphum* contained camphor (32%), geranial (17%), neral (12%). The antibacterial screening of each EO revealed that only the leaf EO of *C. damhaensis* failed to inhibit the growth of all bacteria, while the leaf EO of *C. tonkinense* failed to inhibit the growth of all Gram-negative bacteria. The rest of the EOs exerted antimicrobial activity at the MIC value ranging between 16–256 µg/mL as discussed below. Interestingly, at MIC of 16 µg/mL, only the EO of *C. ovatum* stems demonstrated an effective antibacterial activity against *P. aeruginosa*. At the MIC of 32 µg/mL, the EO of *C. tonkinense* and *C. polyadelphum* leaves inhibited the growth of *E. faecalis*. In comparison, streptomycin was effective against *E. faecalis* and *E. coli*. On the other hand, at the MIC of 64 µg/mL, i) both EOs of C. *ovatum* leaves and stems were effective against *E. faecalis*, *S. aureus*, *E. coli,* and *S. entrica,* while the latter EO was effective against *B. cereus*. In addition, the EO of *C. longipetiolatum* leaves was effective only against *E. faecalis,* whereas the EO of *C. polyadelphum* leaves was effective against *S. aureus* and *B. cereus*. In comparison, only streptomycin exerted an antibacterial activity against *B. cereus* and *S. enterica* at 64 µg/mL. With regards to the determined IC_{50}, the EO of *C. tonkinense* demonstrated the most effective inhibition against the *E. faecalis* activity with the recorded value of 5.7 µg/mL followed by the EO of *C. polyadelphum*, *C. longipetiolatum* and both the leaves and stems EOs of *C. ovatum*, which recorded the value ranging between 10.7–32.4 µg/mL. As for *S. aureus*, the EO of *C. polyadelphum* leaves followed the EOs of *C. ovatum* leaves and then stems, *C. longipetiolatum* leaves and *C. tonkinense* leaves caused the inhibitory growth at the IC_{50} of 24.8, 32.3, 32.8, 36.8 and 37.8 µg/mL, respectively. For B. cereus, the EOs of *C. polyadelphum* leaves followed by *C. ovatum* stems caused the inhibition of growth at the IC_{50} of 30.2 and 33.6 µg/mL, respectively, followed by the leaves of *C. tonkinense*, *C. longipetiolatum* and *C. ovatum* with the IC_{50} ranging from 56.7–65.5 µg/mL. Only the EO of *C. damhaensis* leaves failed to affect the growth of all Gram-positive bacteria. As for the Gram-negative bacteria, the EOs of *C. tonkinense* and *C. damhaensis* leaves failed to produce the IC_{50} against all of them. Furthermore, the EOs of *C. ovatum* leaves and stems were more effective in inhibiting the growth of *E. coli, P. aeruginosa* and *S. enterica* as indicated by the lower IC_{50} value of 32.6 and 32.6; 65.4 and 8, and; 33.2 and 31.2 µg/mL compared to the EO of *C. longipetiolatum* leaves, which were 100.3, 105.7 and 56.8 µg/mL, respectively. Furthermore, the EO of *C. polydelphum* leaves was only effective against *S. enterica* with the IC_{50} of 57.5 µg/mL.

Jusoh et al. (2020) investigated the antibacterial potential of EOs of *Alpinia malaccensis* from the fresh leaves and pseudo stems, collected from the conservation garden Universiti Putra Malaysia, Serdang, Selangor, Malaysia. Following the hydrodistillation process, the obtained EOs were analyzed using the GC-MS and phytoconsitutents identification was carried out using standard procedures. The antibacterial study of the leaves- and pseudo stems-derived

EOs of *A. malaccensis* was conducted using the disc diffusion method against five bacteria of which two were Gram-positive bacteria (i.e., *S. aureus* ATCC29737 and *B. subtilis* ATCC6633) and three were Gram-negative bacteria (i.e., *P. aeruginosa* ATTC9027, *E. coli* ATTC10536 and Pseodomonas putida ATCC49128). Each EO dissolved in the DMSOto the concentration of 4 mg/mL was tested against those bacteria with the DMSO alone used as the negative control and streptomysin sulphate (10 µg/mL) used as the reference antibacterial drug. Following the screening procedure, the MIC value for both EOs, prepared at the concentration of 1 mg/mL, against these bacteria was determined using the microdilution method. Each EO stock solution then underwent a two-fold serial dilution to achieve the concentrations of 1000, 500, 250, 125, 62.5, 31.3, 15.63 and 7.81 µg/mL, which were then tested in the microdilution assay with streptomysin sulphate used as the reference drug. Following the GC-MS analysis, 73 volatile compounds, which constituted 100.0% of the total EO content, were identified and further classified in 47.0% monoterpenes hydrocarbons, 25.2% oxygenated monoterpenes, 13.4% sesquiterpene hydrocarbons, 6.1% oxygenated sesquiterpenes and 8.3% esters. Specifically, the most abundant components of leaves-derived EO of *A. malaccensis* were β-pinene (20.0%), 1,8-cineole (16.5%), trans-caryophyllene (8.7%), α-pinene (6.6%) and *p*-cymene (5.6%). On the other hand, 24 volatile compounds, accounting for 96.0% of the total EO content, were identified and constituted 30.2% monoterpenes hydrocarbons, 49.8% oxygenated monoterpenes, 13.2% sesquiterpene hydrocarbons, 1.9% oxygenated sesquiterpenes and 0.9% esters. Further analysis revealed the major presence of 1,8-cineole (36.81%), β-pinene (13.95%), trans-caryophyllene (7.49%), α-terpineol (6.9%) and α-phellandrene (5.3%). With regards to the antibacterial activity, the pseudo stems-derived EO of *A. malaccensis* exerted a growth inhibitory activity against all bacteria tested with the recorded zone of inhibition range of 14.0–16.0 mm, whereas the leaves-derived EO, which was ineffective against *B. subtilis* and *P. aeruginosa*, exerted growth inhibitory against the rest of the bacteria in the range of 14.0–16.7 mm. In comparison, streptomysin sulphate also inhibited the growth of all bacteria with the recorded zone of inhibition of 15.0–18.7 mm. In terms of the MFC value, the leaves-derived EO of *A. malaccensis* demonstrated strong antibacterial activity against *S. aureus* and *E. coli* followed by P. putida without any activity against *P. aeruginosa* with the recorded value of 7.8, 15.6 and 125.0 µg/mL, respectively. On the other hand, the pseudo stems-derived EO was less effective than the leaves EO, despite inhibiting the growth of all bacteria, with the recorded MCF value ranging between 31.3–125.0 µg/mL.

In a recent study, Donadu et al. (2020) investigated the antibacterial potential of EOs of *Hornstedtia bella Škorničk* leaves, rhizomes and the whole plant collected from Luoi District, Thua Thien Hue Province, Vietnam. Following the hydrodistillation process, the obtained EOs were analyzed using GC-MS and the phytoconsitutents present in the EOs were identified. The antibacterial study of the leaves, rhizomes and whole plant-derived EOs of *H. bella* was performed using the broth dilution method to determine the MIC and Minimum Lethal Concentration (MLC) *values* against 13 microbial strains, which include three reference strains and 10 clinical isolates as: (i) Gram-positive strains

(*S. aureus*, methicillin-resistant *S. aureus*, *S. epidermidis* and *Enterococcus faecalis*); (ii) Gram-negative strains (*E. coli, P. aeruginosa* and *K. pneumonia)*. However, no clear information was given regarding the source(s) of all strains of bacteria. The EOs were diluted to different concentrations from 16 to 0.06% (*v/v*). The solvent used to dissolve the EOs was not mentioned and there was no reference drug used in the study. However, each assay was carried out with a negative sterility control containing an uninoculated broth and a positive growth control that consisted of organisms in the broth. Following the GC-MS analysis, the whole plant EO contained 26 constituents, representing 96.2% of the total oil content. The whole plant EO consisted of monoterpene hydrocarbons (36.98%), sesquiterpene hydrocarbons (31.80%), oxygenated sesquiterpenes (15.95%), and oxygenated monoterpenes (11.47%) as the main classes of compounds. This oil was found to be rich in β-pinene (25.52%). The other constituents included 1,8-cineole (10.50%), α-pinene (8.75%), *E*-β-caryophyllene (5.56%), α-humulene (5.64%), germacrene D (5.84%) and ledene (5.11%). As for the leaves of the EO of *H. bella*, the main classes of compounds reported were sesquiterpene hydrocarbons (69.72%), oxygenated sesquiterpenes (18.82%), and monoterpene hydrocarbons (7.27%), while the main constituents present in the EO were germacrene D (17.47%), viridiflorene (11.13%), *E*-β-caryophyllene (7.32%) and α-humulene (6.48%). A total of 56 compounds were identified in the EO of rhizomes, accounting for 95.25% of its total content. The major classes of compounds present in this EO were sesquiterpene hydrocarbons (32.55%), monoterpene hydrocarbons (28.79%) and oxygenated sesquiterpenes (24.21%). The main constituents found in rhizomes differed completely as compared to the ones found in the leaves, where the rhizome EO was found to be abundant in β-pinene (16.07%), α-humulene (9.68%), β-selinene (7.11%), and epiglobulol (6.48%). The only diterpene found in the leaves was retinal (2.14%). With regards to the antibacterial activity, the EO of rhizomes exerted better antibacterial activity as compared to the leaves EO. The *H. bella* leaves EO exhibited strong inhibition against *S. aureus,* MRSA and *S. epidermidis* with MIC of 2% (*v/v*) and MLC of 4% (*v/v*), followed by 8% (*v/v*) against *E. faecalis,* while the *H. bella* rhizomes EO showed the highest inhibition against *S. aureus*, MRSA and *S. epidermidis* with MIC of 1% (*v/v*) and MLC of 2% (*v/v*), followed by MIC of 4% and MLC of 8% (*v/v*) in *E. faecalis.* In addition, the entire plant's EO exerted inhibition against *S. aureus*, MRSA and *S. epidermidis* with MIC and MLC values from 2 to 4% (*v/v*). All the three EOs recorded MIC and MLC values ranging from 8 to greater than or equal to 16% (*v/v*) in all the Gram-negative strains tested.

Huong et al. (2020b) also reported on the chemical composition and antibacterial activity of Vietnamese Annonaceae plants, namely *Uvaria hamiltonii* Hook. f. & Thoms. and *Fissistigma kwangsiensis* Tsiang & P.T. Li. The leaves of both plants were collected from their natural habitat in an unspecified location in Vietnam. In all, 35 and 51 constituents were identified in the respective EO of *U. hamiltonii* and *F. kwangsiensis* leaves, which account for 94.8 and 98.7% of the volatile contents. Further analysis revealed that the main classes of compounds detected in each EO of *U. hamiltonii* and *F. kwangsiensis* leaves therein included sesquiterpene hydrocarbons (70.0 and 78.7%), oxygenated sesquiterpenes (19.0 and

10.0%) and monoterpene hydrocarbons (5.3 and 10.0%), respectively. In addition, the oxygenated monoterpenes were not identified in *F. kwangsiensis* and present in lower quantity (0.5%) in *U. hamiltonii*. Specifically, the EO of *U. hamiltonii* leaves contained mainly of germacrene D (22.9%), β-caryophyllene (21.1%), bicyclogermacrene (11.2%), caryophyllene oxide (8.6%) and spathulenol (5.8%), whereas the EO of *F. kwangsiensis* consisted mainly of β-caryphyllene (24.5%), δ-cadinene (13.4%), (Z)-β-ocimene (6.7%), α-copaene (5.6%) and γ-muurolene (4.5%). An antibacterial study of *U. hamiltonii* and *F. kwangsiensis* leaves EOs revealed that, overall, the former EO showed effective antibacterial activity towards *B. cereus* (MIC, 5.7 μg/mL) followed by *E. faecalis* (MIC, 8.0 μg/mL), *P. aeruginosa* (MIC, 12.3 μg/mL) and *S. aureus* (MIC, 20.3 μg/mL), while the latter EO displayed the most potent antibacterial action against *P. aeruginosa* (MIC, 3.45 μg/mL) followed by *B. cereus* (MIC, 12.5 μg/mL), *E. faecalis* (MIC, 33.5 μg/mL) and *S. aureus* (MIC, 55.7 μg/mL). Also based on the IC_{50} value, the EO of leaves of *U. hamiltonii* was more effective against *E. faecalis* (IC_{50}, 16.0 vs. 64.0 μg/mL), *B. cereus* (IC_{50}, 64.0 vs. 128.0 μg/mL) and *S. aureus* (IC_{50}, 16.0 vs. 32.0 μg/mL), while *F. kwangsiensis* leaves EO was more effective only against *P. aeruginosa* (IC_{50}, 16.0 vs. 32.0 μg/mL). For comparison, the MIC values for streptomycin were 0.48, 2.07, 1.07, 3.20, 0.28, and 0.97 μg/mL, respectively. Although streptomycin was mentioned as the reference antibacterial drug, it is not clear to which bacteria the MIC values provided above belonged to.

A further attempt by Rozman et al. (2020) investigated the antibacterial property of *Homalomena pineodora* Sulaiman and P.C. Boyce EO against 12 clinical microbial strains was performed using the disc diffusion method. The bacteria tested were believed to be of clinical origin and included four Gram-positive bacteria (i.e., *B. cereus, Bacillus subtilis, S. aureus* and MRSA) and eight Gram-negative bacteria (*E. coli, Proteus mirabilis, Yersinia sp., Klebsiella pneumoniae, Shigella boydii, Salmonella typhimurium, Acinetobacter anitratus* and *P. aeruginosa*). The clinical microbial strains were provided by the Hospital Seberang Jaya, Penang, Malaysia. The results obtained showed that, except for *K. pneumoniae, S. boydii, S. typhimurium*, only nine bacteria growth were inhibited by *H. pineodora* EO with a diameter of growth inhibitory zone ranging between 9.0–11.0 mm in comparison to chloramphenicol that caused an inhibition of a growth zone in the range of 11.0–15.0 mm. Furthermore, the EO exerted an antimicrobial activity against the test microorganisms, with an MIC range of 4.88–78.00 μg/mL. A higher Minimal Lethality Concentration (MLC) was obtained for *E. coli, K. pneumoniae, A. anitratus, S. typhimurium* and *P. aeruginosa* (625.00 μg/mL). The lowest MIC was obtained for *B. cereus, Yersinia* sp., *E. coli, A. anitratus* and *S. typhimurium* (4.88 μg/mL). Despite the findings, no attempt was made to determine the phytocontituents of *H. pineodora* EO using the GC-MS.

Huong et al. (2021) studied the antibacterial potential of the leaves- and stems-derived EOs of *Amomum rubidum* Lamxay & N.S. Lý. The leaves and stems were harvested from mature plants found in Bidoup Nui Ba National Park, Lam Dong Province, Vietnam and underwent hydrodistillation to produce the respective EO. Each EO was then subjected to the GC-MS analysis and the phytoconstituents identification followed by the antibacterial study. The antibacterial activity of the EOs of *A. rubidum* was carried out to determine MIC value using the microdilution broth susceptibility assay against eight standardized ATCC strains from the laboratory stock consisting of two Gram-negative strains (i.e., *E. coli* ATCC25922 and *P. aeruginosa* ATCC25923) and four Gram-positive strains (i.e., *B. subtilis* ATCC1177, *S. aureus* subsp. *aureus* ATCC11632, *A. niger* ATCC9763 and *F. oxysporum* ATCC48112). The authors did not provide further information related to the concentration of EOs tested and the reference antibacterial agents used. The GC-MS analysis identified 30 compounds in the leaves-derived EO that accounted for 99.1% of the total EO content, whereas 49 compounds were identified in the stems-derived EO, which accounted for 97.4% of the total oil content. In terms of the main classes of compounds present in each EO, monoterpene hydrocarbons (52.1%), oxygenated monoterpenes (43.2%), sesquiterpene hydrocarbons (1.5%) and non-terpenes (2.3%), but not oxygenated sesquiterpenes, were detected in the leaves of the EO, while in the stem of the EO, monoterpene hydrocarbons (80.1%), oxygenated monoterpenes (3.1%), sesquiterpene hydrocarbons (10.3%), oxygenated sesquiterpenes (3.5%) and non-terpenes (0.5%) were detected. Specifically, the main constituents of the leaves EO were identified to be 1,8-cineole (37.7%), δ-3-carene (19.5%), limonene (16.3%), α-pinene (3.0%) and myrcene (2.7%), while the EO for the stems, the main constituents were δ-3-carene (21.9%), limonene (17.8%), β-phellandrene (14.6%), α-pinene (5.8%) and α-phellandrene (4.7%). The antibacterial study carried out on the selected Gram-negative and Gram-positive bacteria revealed that the leaves of *A. rubidium* EO displayed stronger antibacterial activity against *P. aeruginosa* with the recorded MIC value of 25 μg/mL. At the MIC value of 50 μg/mL, the leaves-derived EO also inhibited the growth of *F. oxysporum* while the stem-derived EO inhibited the growth of *P. aeruginosa, F. oxysporum* and *C. albicans*. The rest of the bacteria did not produce any inhibitory activity when tested with any of the EOs.

Ginting et al. (2021) investigated the antibacterial activity of EOs extracted from *Syzgium aromaticum* and *Cinnamomum burmannii* against three strains of bacteria, namely *E. coli* ATCC25922 and Extended-Spectrum β-Lactamase (ESBL)-producing *E. coli* from chicken meat (cultured and collected from the Veterinary Public Health Laboratory, Faculty of Veterinary Medicine, Gadjah Mada University, Indonesia) and *Klebsiella pneumoniae* ATCC700603. The *S. aromaticum* bud, obtained from Jogjakarta, Indonesia, and *C. burmannii* bark, procured from Central Java, Indonesia, were subjected to steam distillation to obtain the respective EO, which was then subjected to GC-MS analysis. From the volatile constituent analysis of both EOs, *S. aromaticum* EO revealed the presence of eugenol (82%) and eugenyl acetate (10.5%), β-caryophyllene (8.64%), and α-humulene (1.0%), while *C. burmannii* contained cinnamaldehyde (61.1%), naphthalene (8.9%), eucalyptol (3.5%), copaene (3.5%) and IR-α-pinene (1.7%). The antibacterial study showed that both EOs, at all concentrations used (10, 20 and 40%; *v/v*), demonstrates the antibacterial effect against all bacteria with the recorded range of growth inhibition of 15.0–24.3 mm and 15.6–25.3 mm, respectively, whereas chloramphenicol (the dose was not mentioned) produced a growth inhibitory zone of 34.0 mm. Further investigation revealed that the MIC value recorded against all strains tested for *S. aromaticum* EO was 0.078% (*v/v*), while for *C. burmannii*, the recorded MIC value range between 0.039–0.156%. In addition, the MBC value recorded for the former EO was 0.078–0.156% with highest activity seen against *K. pneumoniae* and ESBL-producing *E. coli,* while the MBC value recorded for the latter EO was also in the similar range of 0.078–0.156%, but a high activity was seen only against *E. coli* ATCC25922 and *K. pneumoniae.*

Antifungal Activity of Essential Oils

The importance of medicinal plants from SeAR as sources of antifungal agents is undeniable based on the increasing number of articles published from year to year and is summarized in Table 10.2. In this antifungal section, most of the discussion did not further describe the phytoconsituents of each EO under review as this had been carried out in the earlier antibacterial section.

Ha et al. (2019) also determined the antifungal property of a Vietnamese plant, *Magnolia hypolampra* using the agar disc diffusion method. The leaves- and twigs-derived each EO of *M. hypolampra*, prepared in the final concentration ranging between 16384 to 2 µg/mL via two-fold serial dilution, was tested against *C. albicans* ATCC10231 followed by the MFC and IC_{50} values determination using the microdilution broth susceptibility assay. The antifungal screening of both EOs of *M. hypolampra* demonstrated that each EO possesses a strong growth inhibitory effect against *C. albicans* as indicated by the growth inhibitory zones that is more than 14.0 mm. The twigs-derived EO exerted a stronger activity against the fungi with the recorded zone of inhibition of 46.0±1.4 mm, whereas for the leaves-derived EO, the recorded zone of inhibition was 27.5±2.1 mm. Additionally, the MFC value of the leaves- and twigs-derived EOs was 4.1 mg/mL against *C. albicans.* Even the IC_{50} value comparison between the leaf- and twig-derived EO revealed that for *C. albicans*, the latter EO produced the lowest IC_{50} value (IC_{50} = 1.6 mg/mL) compared to the former EO (IC_{50} = 2.5 mg/mL). For comparison, nystatin, as the reference antifungal drug, was cited to have MFC value in the range of 2.8 to 5.0 µg/mL, respectively.

Huong et al. (2020a) studied the antifungal potential of the EO of *Zingiber zerumbet* rhizomes against two strains of fungi (*C. albicans* ATCC10231 and *Saccharomyces cerevisiae* ATCC16404) which were used. The articles directly reported on the attempt to determine MFC value of *Z. zerumbet* using the microdilution broth susceptibility assay without mentioning anything about the antimicrobial screening procedure using the disc diffusion method, which was performed in other studies reported earlier. Other than reporting the use of sterile distilled water and DMSO as the positive control, missing information about the reference antifungal drug used was also noted. Following the antifungal study to determine the MFC value against *C. albicans*, EO of the rhizomes of *Z. zerumbet* was found to show no inhibitory effect against the tested fungi. Another issue detected in the article is the classification of *Fusarium oxysporum* ATCC48112 was not clear as in the present form it can be read that the microbe

Table 10.2 Summary of phytoconstituents of EOs from various plants obtained from SeAR that exerted antifungal activity

Fungi	**Plant species**	**Plant Family**	**Parts used**	**Location/ country**	**Phytoconstituents**	**References**
Aspergillus fumigatus	*Piper betle* (betel leaf)	*Piperaceae*	Leaves	Vietnam	ND	Tao et al. 2020
	Cymbopogon flexuosus lemongrass)	Grasses				
	Mentha arvensis (peppermint)	Mints				
	Citrus aurantifolia	*Rutaceae*	Fruits	Songkla, Thailand		
Candida albicans	*Melaleuca cajuputi* Powell (*M. cajuputi*)	Myrtaceae	leaves	Community Forest in Tha Chang District (The Chaipattana Foundation), Chanthaburi Province, Thailand	1,8-naphthyridine derivatives (10.46%), alpha-pyrone (10.11%), terpinolene (9.26%), gamma-terpinene (8.00%), beta-caryophyllene (6.36%), and beta-elemene (5.09%).	Keereedach et al. 2020
	Zingiber cornubracteatum		Leaf	North central, Vietnam	α-pinene (2.7–10.1%) and β-pinene (18.8–67.3%), (*E*)-caryophyllene (1.8–13.9%), germacrene D (0.7–13.7%), bicyclogermacrene (2.7–18.9%), (*E*)-nerolidol (0.9–23.0%).	Huong et al. 2020
			rhizome	North central, Vietnam	α-pinene (8.2–14.5%), β-pinene (8.8–33.1%), limonene (1.0–5.1%), 1,8-cineole (2.5–10.4%), and linalool (0.4–31.0%).	
	Z. cassumunar	Zingiberaceae	rhizome	Sabak Bernam, Selangor, Malaysia.	terpinen- 4-ol (37.7%), b-pinene (20.8%), (*E*)-1-(3,4-dimethoxyphenyl)but-1-ene (13.3%), g-terpinene (5.1%) and (*E*)-1-(3,4-dimethoxyphenyl) butadiene (4.7%).	

Fungi	Plant species	Plant Family	Parts used	Location/ country	Phytoconstituents	References
	Z. ottensi	Zingiberaceae	rhizome	North central, Vietnam	sabinene (21.6%), β-pinene (11.7%), γ-terpinene (5.5%), terpinen-4-ol (17.1%), sesquiterpenoid zerumbone (12.5%).	
			leaf	North central, Vietnam	(*E*)-caryophyllene (28.0%) and trans-β-elemene (17.0%), monoterpene β-pinene (17.1%).	
	Z. nudicarpum	Zingiberaceae	leaf		α-pinene (5.0–10.9%), β-pinene (0.7–34.0%), sesquiterpene hydrocarbons (E)-caryophyllene (6.4–24.3%), α-humulene (2.1–6.4%), germacrene D (0.6–6.5%), and bicyclogermacrene (3.3–16.1%).	
			rhizome		perillene (51.3%), neral (12.3%), and geranial (17.0%).	
	Z. recurvatum	Zingiberaceae	leaf		monoterpenes α-pinene (16.3%) and β-pinene (71.6%).	
			rhizome		(*E*)-caryophyllene (11.3%), bornyl acetate (10.4%), α-humulene (6.9%), and bicyclogermacrene (5.1%).	
	Z. vuquangensis	Zingiberaceae	leaf	North central, Vietnam	α-pinene (11.3%), β-pinene (38.5%), sesquiterpene hydrocarbons trans-β-elemene (5.9%), and (*E*)-caryophyllene (12.2%).	

Table 10.2 contd....

...Table 10.2 contd.

Fungi	Plant species	Plant Family	Parts used	Location/ country	Phytoconstituents	References
	Cinnamomum (*C. damhaensis*)	Lauraceae	leaf	North Central, Vietnam	The leaf essential oil of *C. tonkinense*, rich in β-phellandrene (23.1%) and linalool (32.2%). The major components of the leaf and stem essential oils were eugenol (70.5% and 71.2%, respectively), eugenyl acetate (9.5% and 9.3%, respectively) and linalool (5.9% and 8.3%, respectively)	Dai et al. 2020
	Piper betle (betel leaf)	*Piperaceae*	Leaves	Vietnam	ND	Tao et al. 2020
	Cymbopogon lexuosus (lemongrass)	Grasses				–do–
	Mentha arvensis (peppermint)	Mints				–do–

is categorized as a Gram-positive strain, whereas it had earlier been classified as fungi. However, regardless of the classification the tested EO did not produce any significant antifungal effect.

An attempt has also been made by Dai et al. (2020) to determine the antifungal activity of EOs from several species of *Cinnamomum*. The leave and stems of *C. ovatum* and the leaves of *C. tonkinense*, *C. damhaenensis*, *C. longipetiolatum* and *C. polyadelphum* were hydrodistillated to obtain the respective EO, which was screened for the antifungal potential against *C. albicans* ATCC10231. Then, the MFC and IC_{50} values of each EO, prepared to the final concentration ranging between 16384 to 2 μg/mL, was determined using the microdilution broth susceptibility assay. Nystatin and cycloheximide were used as reference antifungal standards, whereas the positive control received only the serial dilutions of the sample without microorganisms and the negative control received sterile distilled water and medium. With regards to the antifungal screening performed on each EOs, the EO of *C. ovatum* stems and *C. tonkinense* leaves inhibited the growth of *C. albicans* with the recorded MCF value of 32 μg/mL. In comparison, cyclohexamide was effective only against *C. albicans* with the recorded MCF value of 32 μg/mL. On the other hand, at the MCF of 64 μg/mL, the EO of C. *ovatum* leaves were effective against *C. albicans.* With regards to the determination of IC_{50} value, the EOs of *C. ovatum* stems and *C. tonkinense* leaves effectively inhibited the growth of *C. albicans* at the IC_{50} of 15.7 μg/mL followed by the EOs of *C. ovatum* leaves, which inhibited the growth at the IC_{50} of 33.2 μg/mL. On the other hand, the EOs of *C. longipetiolatum* and *C. polydelphum* leaves inhibited the growth of *C. albicans* only at the IC_{50} of 112.5 and 123.5 μg/mL, respectively, while the EO of *C. damhaensis* leaves failed to produce any effect.

Jusoh et al. (2020) also investigated the antifungal activity of *Alpinia malaccensis* EOs extracted from its fresh leaves and pseudo stems, that was collected in Malaysia. Both EOs, at the concentration of 4 mg/mL, were tested against a yeast strain, *C. albicans* ATCC10231, and a fungi strain, *A. niger* ATCC16888, using the disc diffusion method with nystatin (100 IU) used as the reference antifungal drug. The MFC value determination was also performed on both EOs, prepared in the concentration range of 1000–7.81 μg/mL via the two-fold serial dilution of the 1 mg/mL stock solution, using the microdilution method with nystatin used as the reference drug. Following the screening procedure, only the pseudo stems-derived EO of *A. malaccensis* inhibited the growth of both *C. albicans* and *A. niger* with the recorded zone of inhibition of 14.3 and 15.0, whereas the leaves-derived EO only inhibited the growth of *C. albicans* (14.4 mm). In addition, the MCF value recorded for both bacteria treated with pseudo stems and for *A. niger* treated with the leaves EO was 125.0 μg/mL. Nystatin also inhibited the growth of both *C. albicans* and *A. niger* with the similar zone of inhibition recorded (15.0 mm) and produced the MCF value of 7.8 μg/mL against both microbes.

In a recent study, Donadu et al. (2020) investigated the antifungal potential of leaves, rhizomes and whole plant-derived EOs of *H. bella*. The plant parts were hydrodistillated to obtain the respective EO, which was screened for the antifungal potential against *C. albicans*, *Candida glabrata*, *Candida tropicalis and Candida parapsilosis*. The broth dilution method was used to determine the MIC and MLC values. The EOs were diluted to different concentrations from 16 to 0.06% (*v/v*). There was no reference drug used in the study. Each assay was carried out with a positive-growth control that consists of organisms in the broth and a negative sterility control containing the uninoculated broth. With regards to the antifungal screening performed on each EOs, leaves and rhizomes EOs of *H. bella* indicated the best antifungal activity against *C. tropicalis* and *C. parapsilosis*. The leaves and rhizomes EO of *H. bella* displayed strong inhibition against *C. tropicalis* with MIC of 2% (*v/v*) for both and MLC of 4% (*v/v*) and 2% (*v/v*), respectively. In addition, the same EO exhibited inhibition against *C. parapsilosis* with MIC and MLC values of 4 (*v/v*) and 2% (*v/v*), respectively. Meanwhile, inhibition against *C. albicans* and *C. glabrata* were found to be less effective with the MIC and MLC values ranging from 8 to greater than or equal to 16% (*v/v*). The whole plant EO was found to be the least effective in inhibiting all the *Candida* sp. tested as the MIC and MLC values were 16% (*v/v*) for all the strains except *C. albicans*, resulted in MLC value greater than 16% (*v/v*).

Huong et al. (2020b) also determined the antifungal activity of Annonaceae plants, namely *Uvaria hamiltonii* and *Fissistigma kwangsiensis* that were obtained in Vietnam, against *C. albicans* ATCC10231. The leaves-derived EOs of both plants were found to exert antifungal activity against *C. albicans* with *F. kwangsiensis* EO demonstrating stronger an anti-candidal effect indicated by the lower MFC (16.5 μg/mL) and IC_{50} (32.0 μg/mL) when compared against *U. hamiltonii* EO, which produced the anti-candidal effect of 32.6 and 64.0 μg/mL, respectively. Comparatively, nystatin and cylohexamide produced MFC value of 8.0 and 3.20 μg/mL, respectively. The obtained MIC and IC_{50} values are an indication that the leaves-derived EOs of *U. hamiltonii* and *F. kwangsiensis* possessed potent anti-candidal activity against the tested microbe.

Keereedach et al. (2020) reported on the antifungal activity of leaf- and twig-derived EO of *Melaleuca cajuputi* Powell against several fluconazole-resistant clinical strains of *C. albicans*. In the study, fresh leaves, and twigs of *M. cajuputi* were collected from the Community Forest in Tha Chang District, Chanthaburi Province, Thailand and subjected to hydrodistillation to obtain the respective EO. The EOs were subjected to the GC-MS analysis and antifungal study. Following the GC-MS analysis, five major constituents were identified, namely 1,8-naphthyridine derivatives (10.5%), α-pyrone (10.1%), terpinolene (9.3%), γ-terpinene (8.0%) and β-caryophyllene (6.4%). In antifungal study, the broth macrodilution method was applied while 16 fluconazole-resistant *C. albicans* clinical strains (labelled as S7/1, S8/1, S9/1, S10/1, S14/1, S15/1, U6/1, U6/2, U7/1, U8/1, U11/1, U15/1, U15/2, U19/1, U21/1 and U25/1), came from the collection of the Department of Microbiology, Faculty of Science, Burapha University, Chonburi Province, Thailand and *C. albicans* ATCC90028 were used. These *C. albicans* strains were earlier subjected to genomic DNA screening for efflux-pump genes, namely CDR1, CDR2, and MDR1, which revealed the distribution of efflux-pump genes only in 14 resistant strains (87.5%) in three distinct patterns, namely pattern A was CDR1-, CDR2-, and MDR1-positive (detected in S7/1, S8/1, S9/1, S10/1, S14/1, S15/1, U6/1, U6/2, U7/1 and U25/1 strains), pattern B was CDR1- and

MDR1-positive (detected only in U15/1strain) and pattern C was only MDR1-positive (detected in U8/1, U11/1 and U21/1). Without reporting the antifungal screening findings, the authors proceeded with the determination of MIC and MFC values of both *M. cajuputi* EOs against the 14 resistant strains and with *C. albicans* ATCC90028 as the quality control strain. The final concentrations of each EO ranging from 0.04 to 5 µl/mL (0.04, 0.08, 0.16, 0.31, 0.63, 1.25, 2.5, and 5 µl/mL) was prepared via a two-fold serial dilution and tested. In addition, the MFC/MIC ratio was calculated to determine antifungal activity whereby a substance with MFC/MIC ratio ≥4 was defined as having a fungistatic activity while a substance with the MFC/MIC ratio <4 was defined as having a fungicidal activity. The results obtained shows that the EOs of *M. cajuputi* possess an antifungal activity against all resistant strains with the recorded range of MICs and MFCs from 0.31 to 1.25 µL/mL and 0.63 to 1.25 µL/mL, respectively. Specifically, for all clinical strains that exert patten A distribution of efflux-pump genes, as well as one strain with pattern B distribution, the MIC and MFC value recorded for each strain was 1.25 and 1.25 µL/mL, whereas for strains that showed the pattern C distribution, the MIC and MFC values fell in the range of 0.31 to 1.25 and 0.63 to 1.25 µL/mL, respectively. Following the MFC/MIC ratio calculation and based on the score used whereby the ratios fall within the range of 1 to 2, the EOs of *M. cajuputi* is classified to exert an antifungal activity against all clinical strains of *C. albicans* through a fungicidal activity.

A study on the antifungal activity of EO derived from several Zingiberaceae plants against *C. albicans* was reported by Huong et al. (2020c). The Zingiberaceae plants, namely *Z. cornubracteatum, Z. nudicarpum, Z. neotruncatum, Z. nitens, Z. ottensii, Z. recurvatum* and *Z. vuquangensis*, were used in this study. Their leaves, stems, or rhizomes were acquired from north-central Vietnam and subjected to hydrodistillation to extract the respective EOs. Each EO compositions were profiled using GC-MS and GC-FID. The rhizome EOs of *Z. cornubracteatum* were mainly composed of monoterpene hydrocarbons and oxygenated monoterpenoids. The major components were α-pinene (8.2–14.5%), β-pinene (8.8–33.1%), limonene (1.0–5.1%), 1,8-cineole (2.5–10.4%), and linalool (0.4–31.0%). As for the leaves the part of *Z. cornubracteatum,* both α-pinene and β-pinene were the major components (2.7–10.1% and 18.8–67.3%, respectively). In addition, sesquiterpene hydrocarbons, including (*E*)-caryophyllene (1.8–13.9%), germacrene D (0.7–13.7%), bicyclogermacrene (2.7–18.9%), as well as the sesquiterpenoid (*E*)-nerolidol (0.9–23.0%), were also abundant in the leaves EOs. The phytochemistry analysis also found that the leaves EOs of *Z. nudicarpum* were rich in α-pinene (5.0–10.9%), β-pinene (0.7–34.0%), (*E*)-caryophyllene (6.4–24.3%), α-humulene (2.1–6.4%), germacrene D (0.6–6.5%), and bicyclogermacrene (3.3–16.1%). As for *Z. neotruncatum*, the most abundant EOs were in the rhizome with perillene (51.3%), neral (12.3%), and geranial (17.0%). Camphene (40.4%) dominated the rhizome EO composition of *Z. nitens*, followed by bornyl acetate (14.5%), (*E*)-β-ocimene (12.7%), and α-pinene (10.5%). In contrast, the leaves EO of *Z. nitens* was composed largely of α-zingiberene (17.4%), α-pinene (11.2%), β-sesquiphellandrene (10.1%), (*E*)-nerolidol (10.0%), zingiberenol (7.2%), β-pinene (6.0%), and *ar*-curcumene (5.2%). While for the EO from *Z. ottensii*, the leaves part showed abundant of sesquiterpene hydrocarbons, including (*E*)-caryophyllene (28.0%) and *trans*-β-elemene (17.0%), along with the monoterpene β-pinene (17.1%), the rhizome EO contains monoterpenes as the major components, such as sabinene (21.6%), β-pinene (11.7%), γ-terpinene (5.5%), and terpinen-4-ol (17.1%). Another Zingiber plant, *Z. recurvatum,* showed an abundance of monoterpenes α-pinene (16.3%) and β-pinene (71.6%) in the leaves EO, while the major components in the rhizome EO were (*E*)-caryophyllene (11.3%), bornyl acetate (10.4%), α-humulene (6.9%), and bicyclogermacrene (5.1%). For *Zingiber vuquangensis*, the leaves and rhizome EOs were both rich in α-pinene (11.3 and 9.8%, respectively) and β-pinene (38.5 and 45.0%, respectively). The sesquiterpene hydrocarbons *trans*-β-elemene (5.9 and 10.0%), and (*E*)-caryophyllene (12.2 and 14.4%) were major components in the leaves and stem EOs, respectively. The antifungal activity of Zingiber EOs was carried out on the pathogenic yeast *Candida albicans* (ATCC10231). The results from the antifungal assays suggested that *Candida albicans* was relatively sensitive to the *Zingiber* EOs. The EO with the best overall antifungal activity was from *Z. nudicarpum* rhizome with MIC < 10 μg/mL against *C. albicans*. The leaves EO of *Z. recurvatum* and the leaves and stem EOs of *Z. cornubracteatum* rich in α-pinene (16.3, 10.1, and 9.9%, respectively) and β-pinene (71.6, 67.3, and 66.8%), but these EOs showed significantly lower antimicrobial activity. The presence of α-pinene and β-pinene cannot explain the antimicrobial activities of *Zingiber* EOs, and synergistic or antagonistic interactions which likely contribute to it. Nevertheless, several *Zingiber* EOs have shown excellent antimicrobial activity and should be investigated further for controlling Gram-positive bacterial and yeast infections.

Huong et al. (2021) studied the antifungal potential of EOs of *A. rubidum* leaves and stems against two strains of yeast, namely *C. albicans* ATCC10231 and *S. cerevisiae* ATCC16404 by determining their MCF value using the microdilution broth susceptibility assay. However, no information about the reference antifungal agent as well as the concentration of EO used was provided. Following the study, only the stems-derived EO of *A. rubidium* exerted an antifungal activity by only inhibiting the growth of *C. albicans* at the recorded MFC value of 50 μg/mL.

Discussion and Conclusion

In the present chapter, a continuous exploration on the potential of EOs as antiparasitic, antibacterial and antifungal agents is evident by the increase number of articles published in the last 3 yr. EOs possess important volatile compounds in rather high concentrations (20–70%) in comparison to other trace components with diverse bioactivities including an antimicrobial potential. Most of the anti-infective activities observed were further supported by the phytoconstituents analysis using GC, GC-MS, and GC-FID, which revealed the presence of antimicrobial compounds at a high concentration in each EO studied. These volatile compounds, chemically derived from terpenes and their oxygenated derivatives (i.e., terpinen-4-ol, 1,8-cineole, α-pinene, camphor, thymol, γ-terpinene, and linalool), constitute the major content of EOs and might help to justify a variable range of anti-microbial activity against microbes. Due to their hydrophobic characteristics, these types of compounds were able to partition with the lipid component of the cell membrane of bacteria altering the membrane permeability, which results in leakage of critical molecules and ions from the bacterial cell and, eventually, the death of bacterial cell (Chouhan et al. 2017). This could help explain why EOs were more effective against Gram-positive bacteria rather than Gram-negative bacteria.

Although the antimicrobial activity is well established, in most cases, the real effect of EOs was significantly weaker compared to the reference drugs as indicated by the high MIC value recorded which could be attributed to the high volatility of the phytoconstituents, which limits the effective time of action (Abers et al. 2021). To overcome this problem and ensure that the promising antimicrobial activity of EOs are fully utilized, researchers have applied a technique known as nano-encapsulation, which simultaneously increases the antimicrobial potency of essential oils by controlled/sustained release and facilitating close interaction with the microorganisms (Chouhan et al. 2017). Besides, procedures exploiting the synergies between essential oils, constituents of essential oils, and antibiotics along with essential oils have also been recommended. Despite this issue, there are EOs that did possess a high antimicrobial potential as reported by Dai et al. (2020), Jusoh et al. (2020), Huong et al. (2020b), Keereedach et al. (2020) just to name a few, which warrant further in-depth study in order to elucidate the mechanisms of action involved for future anti-infective drug development. As there is a lack of detailed knowledge about the mechanism of the individual EO components, parallel efforts to screen for an antimicrobial agent should also be used in the future to explore the mechanism of individual EO components, along with an initiation in systematically investigating the synergy mechanisms among different components (Chouhan et al. 2017). In conclusion, this chapter presents the current knowledge about researches related to the potential of EO as antimicrobial agents carried out within SeAR and highlight some of the promising EOs that warrant further investigation.

Acknowledgements

Zainul Amiruddin Zakaria thanks the Faculty of Medicine and Health Sciences, Universiti Malaysia Sabah, Sabah, Malaysia for the technical support.

References

Abers, M., S. Schroeder, L. Goelz, A. Sulser, T. St. Rose, K. Puchalski, et al. 2021. Antimicrobial activity of the volatile substances from essential oils. BMC Complement. Med. Ther. 21: 124. https://doi.org/10.1186/s12906-021-03285-3

Alviano, D.S. and C.S. Alviano. 2009. Plant extracts: search for new alternatives to treat microbial diseases. Curr. Pharm. Biotechnol. 10: 106–121. https://doi.org/10.2174/138920109787048607

Astutik, S., J. Pretzsch and J.N. Kimengsi. 2019. Asian medicinal plants' production and utilization potentials: A review. Sustainability. 11: 1–33. https://doi:10.3390/su11195483

Chomchalow, N. 2002. Production of aromatic plants in Asia: An overview. AU J. Technol. 6: 43–53.

Chouhan, S., K. Sharma and S. Guleria. 2017. Antimicrobial activity of some essential oils—Present status and future perspectives. Medicines. 4: 58. https://doi.org/10.3390/medicines4030058

Dai, D.N., N.T. Chung, L.T. Huong, N.H. Hung, D. Chau, N.T. Yen, et al. 2020. Chemical compositions, mosquito larvicidal and antimicrobial activities of essential oils from five species of *Cinnamomum* growing wild in North Central Vietnam. Molecules. 25: 1303.

Dawood, M.A.O., M.F. ElBasuini, A.I. Zaineldin, S. Yilmaz, M.T. Hasan, E. Ahmadifar, et al. 2021. Antiparasitic and antibacterial functionality of essential oils: An alternative approach for sustainable aquaculture. Pathogens. 10: 185. https://doi.org/10.3390/pathogens10020185

Donadu, M.G., N.T. Le, D.V. Ho, T.Q. Doan, A.T. Le, A. Raal, et al. 2020. Phytochemical compositions and biological activities of essential oils from the leaves, rhizomes and whole plant of *Hornstedtia bella* Škorničk. Antibiotics. 9: 334.

Ginting, E.V., E. Retnaningrum and D.A. Widiasih. 2021. Antibacterial activity of clove (*Syzygium aromaticum*) and cinnamon (*Cinnamomum burmannii*) essential oil against extended-spectrum β-lactamase-producing bacteria. Vet. World. 14: 2206–2211.

Gulati, B.K., S. Sharma and M.V. Vardhana Rao. 2021. Analyzing the changes in certain infectious and parasitic diseases in urban population of India by using medical certification of cause of death data. Indian J. Community. Med. 46: 20–23. https://doi.org/10.4103/ijcm.IJCM_77_20

Ha, C.T., T.H. Thai, N.T. Hien, H.T. Anh, L.N. Diep, D.T. Thuy, et al. 2019. Chemical composition and antimicrobial activity of the leaf and twig essential oils of *Magnolia hypolampra* growing in Na Hang Nature Reserve, Tuyen Quang Province of Vietnam. Nat. Prod. Commun. 14: 1–7.

Huong, L.T., V.H. Chinh, N.T. An, N.T. Viet, N.H. Hung, N.T. Thuong, et al. 2020a. *Zingiber zerumbet* rhizome essential oil: Chemical composition, antimicrobial and mosquito larvicidal activities. European J. Med. Plants. 30: 1–12.

Huong, L.T., N.T. Chung, D.T.M. Chau, D.N. Dai and I.A. Ogunwande. 2020b. Annonaceae essential oils: Antimicrobial and compositions of the leaves of *Uvaria hamiltonii* Hook. f. & Thoms. and *Fissistigma kwangsiensis* Tsiang & P. T. Li. Rec. Nat. Prod. 281: 2108–2161.

Huong, L.T., N.T. Chung, T.T. Huong, L.N. Sam, N.H. Hung, I.A. Ogunwande, et al. 2020c. Essential oils of *Zingiber* Species from Vietnam: Chemical compositions and biological activities. Plants. 9: 1269.

Huong, L.T., N.T. Viet, L.N. Sam, C.N. Giang, N.H. Hung, D.N. Dai, et al. 2021. Antimicrobial activity of the essential oils from the leaves and stems of *Amomum rubidum* Lamxay & N. S. Lý Bol Latinoam Caribe. Plant Med. Aromat. 20: 81–89.

Jusoh, S., H.M. Sirat, F. Ahmad, N. Basar, M.B. Bakar, S. Jamil, et al. 2020. Essential oils of leaves and pseudo stems *Alpinia malaccensis* and antimicrobial activities. J. Phys. Conf. Ser. 1529: 042050.

Karunamoorthi, K., K. Jegajeevanram, J. Vijayalakshmi and E. Mengistie. 2013. Traditional medicinal plants: A source of phytotherapeutic modality in resource-constrained health care settings. J. Evid. Based Complementary Altern. Med. 18: 67–74.

Kebede, T., E. Gadisa and A. Tufa. 2021. Anti-infective activities evaluation and phytochemical screening of some selected medicinal plants: A possible alternative in the treatment of multidrug-resistant microbes. PLoS ONE 16: e0249253. https://doi.org/10.1371/journal.pone.0249253

Keereedach, P., K. Hrimpeng and K. Boonbumrung. 2020. Antifungal activity of Thai cajuput oil and its effect on efflux-pump gene expression in fluconazole-resistant *Candida albicans* clinical isolates. Int. J. Microbiol. 4: 2020.

Le, T.B., C. Beaufay, D.T. Nghiem, M.-P. Mingeot-Leclercq and J. Quetin-Leclercq. 2017. In vitro anti-leishmanial activity of essential oils extracted from Vietnamese plants. Molecules. 22: 1071. doi: 10.3390/molecules22071071

Le, T.B., C. Beaufay, D.T. Nghiem, T.A. Pham, M.P. Mingeot-Leclercq and J. Quetin-Leclercq. 2019. Evaluation of the anti-trypanosomal activity of Vietnamese essential oils, with emphasis on *Curcuma longa* L. and its components. Molecules. 24: 1158. https://doi.org/10.3390/molecules24061158

Middleton, D.J., K. Armstrong, Y. Baba, H. Balslev, K. Chayamarit, R.C.K. Chung, et al. 2019. Progress on Southeast Asia's flora projects. Gard. Bull. Singapore. 71: 267–319. https://doi.org/10.26492/gbs71(2).2019-02

Mustafa, G., R. Arif, A. Atta, S. Sharif and A. Jamil. 2017. Bioactive compounds from medicinal plants and their importance in drug discovery in Pakistan. Mat. Sc. Pharm. 1: 17–26.

Oyen. L.P.A. and X.D. Nguyen. 1999. Plant resources of South-East Asia: No. 19 – Essential-oil plants. pp. 1–277. Backhuys Publishers; Leiden, Netherlands.

Petrovska, B.B. 2012. Historical review of medicinal plants usage. Pharmacogn. Rev. 6: 1–5. https://doi.org/10.4103/0973-7847.95849

Pintong, A.R., J. Ruangsittichai, S. Ampawong, K. Thima, P. Sriwichai, N. Komalamisra, et al. 2020. Efficacy of *Ageratum conyzoides* extracts against *Giardia duodenalis* trophozoites: an experimental study. BMC Complement. Med. Ther. 20: 1–9.

Rozman, N., W.Y. Tong, C.R. Leong, M.R. Anuar, S. Karim, S.K. Ong, et al. 2020. *Homalomena pineodora* essential oil nanoparticle inhibits diabetic wound pathogens. Sci. Rep. 10: 3307.

Shezryna, S., N. Anisah, I. Saleh and R.A. Syamsa. 2020. Acaricidal activity of the essential oils from *Citrus hystrix* (Rutaceae) and *Cymbopogon citratus* (Poaceae) on the cattle tick *Rhipicephalus (Boophilus) microplus* larvae (Acari: Ixodidae). Trop. Biomed. 37: 433–442.

Tao, V., T. Tram, N. Hien, T. Dung and T. Tuan. 2020. The antifungal activity of essential oils from some plants in Vietnam against the pathogenic fungi *Candida albicans* and *Aspergillus fumigatus*. Vietnam J. Sci. Technol. 62: 70–75.

Wink, M. 2012. Medicinal plants: a source of anti-parasitic secondary metabolites. Molecules. 17: 12771–12791. https://doi.org/10.3390/molecules171112771

Wińska, K., W. Mączka, J. Łyczko, M. Grabarczyk, A. Czubaszek and A. Szumny. 2019. Essential oils as anti-infective agents-myth or real alternative?. Molecules. 24: 2130. https://doi.org/10.3390/molecules24112130

Zinner, S.H. 2005. The search for new anti-infectives: why we need new options. Expert Rev. Anti Infect. Ther. 3: 907–913. https://doi.org/10.1586/14787210.3.6.907

Index

For Product Safety Concerns and Information please contact our
EU representative GPSR@taylorandfrancis.com Taylor & Francis
Verlag GmbH, Kaufingerstraße 24, 80331 München, Germany